METHODS IN MOLECULAR BIOLOGY™

Series Editor
John M. Walker
School of Life Sciences
University of Hertfordshire
Hatfield, Hertfordshire, AL10 9AB, UK

For further volumes:
http://www.springer.com/series/7651

Plant Meiosis

Methods and Protocols

Edited by

Wojciech P. Pawlowski

*Department of Plant Breeding and Genetics, Cornell University,
Ithaca, NY, USA*

Mathilde Grelon

Institut National de la Recherche Agronomique (INRA), Centre de Versailles-Grignon, Versailles, France

Susan Armstrong

School of Biosciences, The University of Birmingham, Birmingham,UK

Editors
Wojciech P. Pawlowski
Department of Plant Breeding and Genetics
Cornell University
Ithaca, NY, USA

Susan Armstrong
School of Biosciences
The University of Birmingham
Birmingham, UK

Mathilde Grelon
Institut National de la Recherche
 Agronomique (INRA)
Centre de Versailles-Grignon
Versailles, France

ISSN 1064-3745   ISSN 1940-6029 (electronic)
ISBN 978-1-62703-332-9   ISBN 978-1-62703-333-6 (eBook)
DOI 10.1007/978-1-62703-333-6
Springer New York Heidelberg Dordrecht London

Library of Congress Control Number: 2013933045

Foreword

It is generally acknowledged that meiosis is a key event in the life cycles of sexually reproducing eukaryotes. In addition to providing for the necessary reduction of chromosome number between the diplo- and haplophases of life cycles to compensate for nuclear fusion during fertilization it also, crucially, generates new combinations of chromosomes and genes through recombination. The central importance of meiosis for reproduction, fertility, and genetic variation has underpinned the long history of meiotic studies, and the continuing strong research effort to better understand this fascinating process.

Meiosis research predates the adoption of the term "meiosis" by 20–30 years. The last two decades of the nineteenth century witnessed an increasing interest in reproductive processes, initially in animals and later in plants. During this period, a consensus emerged of the necessity for a "reduction division" and, in parallel, cytological studies, facilitated by improvements in microscopy and development of synthetic stains for chromosomes, gradually laid the foundations for a limited understanding of this process. The key development at this time was the realization that reduction involved two consecutive nuclear divisions, and the process was eventually christened "meiosis" in 1905. The following three decades witnessed an escalation of cytogenetic analysis of meiosis, facilitated by further technical improvements in microscopy, improved stains, and the introduction of chromosome squash techniques that largely replaced earlier sectioning techniques. As a result, the main structural/mechanical events of meiosis had been thoroughly understood and described by 1940. The 1950s onwards saw an acceleration of meiosis research with the introduction of new techniques including autoradiography, microdensitometry, electron microscopy, and new specialized chromosome "banding" methods. At the same time meiosis research was extended to other organismal groups including fungi and especially yeast, which was to achieve such prominence in meiosis research in later years.

The last three decades have witnessed perhaps the most fundamental shift in meiosis research with the application of molecular biology techniques to meiotic processes and phenomena. Our understanding of meiosis has been greatly enhanced by the application of such techniques as fluorescent in situ hybridization, protein immunolocalization, proteomic analysis, and identification and characterization of meiotic genes. It is surprising, however, that the introduction of these new methodologies has not supplanted cytological analysis of meiosis; instead there has been a remarkable synergy between cytology and molecular biology that has been a hallmark of meiosis research in recent years and across a wide range of organisms from fungi through invertebrates to mammals including humans and of course plants.

The focus of this volume is on molecular biology approaches to *plant* meiosis and it is pertinent therefore to consider the more general role and significance of plant meiosis research and its associated benefits and problems. Historically, plants have played a prominent role in meiosis research, especially during what we might term the cytogenetic era. This was partly due to their having certain advantages in terms of ease of cytogenetic analysis (large chromosomes, etc.) and partly due to the economic imperative to better understand influences on fertility and genetic variation in plant breeding programmes, especially where polyploidy is involved.

The great majority of plant meiosis studies, both historically and more recently, have been conducted using male meiocytes, termed pollen mother cells, because they are vastly more numerous and also more accessible than their female counterparts (but see below). Unlike animals, plants lack a predetermined germ line. Instead, sporogenous cells arise from somatic meristematic cells in response to developmental cues. Sporogenous cells proliferate asynchronously by mitotic division in developing anthers until a certain critical cell number is achieved, at which point they arrest before synchronously entering meiosis. Pollen mother cells develop a thick callose wall but at early stages of meiosis the pollen mother cells of an anther locule are connected by cytoplasmic channels (plasmodesmata) so that the meiocytes can be isolated and collected as cylinders of interconnected cells, which constitutes an advantage for some experimental procedures. At later meiotic stages the cytoplasmic connections are severed so that individual cells are effectively isolated from their neighbors.

Synchrony of meiotic development and progression within anther locules also extends to synchrony between locules and, in some cases, to synchrony between anthers of individual flowers. This synchrony confers some advantages. First, it allows the selection of many cells at particular, required stages of meiosis for analysis. This is advantageous for cytological analysis but is particularly relevant in the context of transcriptomic and proteomic analyses (Chapters 20 and 21) where obtaining a sufficient yield of meiocyte extract attributable to given stages can be limiting. Secondly, meiotic synchrony facilitates the analysis of meiotic progression and the timing of defined molecular events by means of BrdU incorporation (Chapters 12 and 14). This is because a large cohort of cells can be simultaneously labelled in the meiotic S-phase and their progression through meiosis can be timed and analyzed.

The callose walls that encapsulate pollen mother cells undoubtedly constitute a hindrance to plant meiosis studies. True, these walls can be enzymatically removed from fixed pollen mother cells for the preparation of chromosome spreads (Chapters 1–4, 9–11, and 13, 14) and in this respect constitute only a minor problem. Live pollen mother cells can be similarly treated to produce protoplasts but with unknown consequences for the normal progression of meiosis. The syncytial organization of pollen mother cells during early meiosis constitutes a related problem since many events of critical importance occur when cells are united in a sausage-like column of cells, not readily separated, that can be 2, 3, or more cells deep. These features add to the difficulty of analyzing meiotic processes in vivo in pollen mother cells and have prompted the application of specialized microscopy capable of resolving chromosome structures and fluorescent signals in intact cells and tissues many micrometers thick (Chapters 6–8). In common with animal systems, female meiosis in plants is more difficult to analyze. This is partly because there are many fewer embryo-sac mother cells compared to pollen mother cells and partly because the embryo-sac mother cells are embedded in a mass of ovular tissue which again can require the application of specialized microscopy (Chapter 5).

The choice of plant species for meiotic research has changed significantly in recent years with the advent of molecular genetic and molecular cytogenetic methods. In earlier decades, plants were chosen for analysis because they had large chromosomes or they had some particular features of interest, e.g., localized chiasmata, or because of their economic importance as crop plants. Ironically, the principal dicotyledenous plant model for meiosis research, *Arabidopsis thaliana*, was chosen for its small and relatively simple genome and hence very small chromosomes that were initially considered to be a challenge for cytological analysis. *Arabidopsis* is a weed species of no direct economic importance but it happens to be a relative of the crop Brassicas, which has fortuitously allowed some aspects of meiotic gene identification and characterization to be carried out in cytologically more tractable situations. Molecular genetic and cytogenetic analysis of meiosis in monocotyledonous plants has been largely directed at cultivated cereal species with relatively small genomes such as rice and maize but because of their economic importance, bread wheat, with a rather large genome, and also barley have figured to some extent. An important approach to the analysis of plant meiosis has been the identification and characterization of meiotic genes and the analysis of mutants that are defective in the expression of these genes. Unlike yeast or mouse, mutations of known plant genes cannot be generated at will by homologous recombination technology. Initially the approach followed in *Arabidopsis* was to identify potential meiotic mutants, based on sterility phenotypes, in collections of T-DNA insertional mutants, followed by cytological analysis to verify a meiotic phenotype and then gene cloning based on proximity to the T-DNA inserts. This approach also required complementation analysis to confirm the association of the cloned gene with a meiotic phenotype. The procedure was extremely laborious and time-consuming but did lead to the identification and characterization of important meiotic genes including *AtASY1* whose role in meiosis is still being investigated after 15 years. Fortunately the sequencing of the *Arabidopsis* genome led to the rapid identification of several meiotic genes by homology to known meiotic genes from other systems. This approach has proved to be robust when dealing with genes (such as *AtSPO11-1, AtSPO11-2, AtRAD51, AtDMC1, AtMSH4/5, AtMER3*) that show strong evolutionary sequence conservation. However there are some genes that show little primary homology to functionally equivalent genes in other organisms although the proteins they encode are structurally similar. The *ZYP1* genes of *Arabidopsis* that encode the central element protein of the synaptonemal complex is just such a case.

Meiotic progression is subject to systems which monitor key processes such as DNA recombination and chromosome orientation on meiosis I and II spindles. Defects in these processes trigger checkpoints, which can lead to delay in progression or to meiotic arrest. In many organisms, from yeasts to mammals, meiotic mutants that are defective in key processes frequently exhibit checkpoint-induced arrest, e.g., at pachytene, leading to cell death by apoptosis. Interestingly, and usefully, most plant meiotic mutants do not arrest (*Atrmi1/Atblap75* is a rare case where arrest at telophase I has been clearly observed). They may, and often do, exhibit delay in meiotic progression but in most cases they are able to progress to completion of meiosis although inevitably the meiotic products are abnormal. This is typified by the *Atmsh4* and *Atmlh3* mutants, which exhibit delays of 8 and 25 h, respectively. This feature of plant meiosis has some advantages since, for example, recombination defects can be analyzed cytologically at metaphase I (chiasmata) or alternatively by genetical/molecular analysis of pollen or progeny (Chapter 18).

Polyploidy is very common among plant species and this raises special issues for the analysis of meiosis in many plants (e.g., wheat), quite apart from the fundamental issues of meiosis. However, plant meiosis research has to contend with a further, related, complication—namely, that many apparently diploid species, exhibiting disomic meiotic behavior, have experienced ancient duplications of their genomes, in some cases repeated duplications at different times in their evolutionary histories. So, for example, there is evidence of two ancient genome duplications in *Arabidopsis* and it is estimated that about 60 % (or more) of the *Arabidopsis* genome is duplicated as a result of one or both of these events. In many cases duplicated genes have undergone silencing but in some well-documented cases they have retained related functional activity, as in the case of *ZYP1a* and *ZYP1b*. This situation can have implications for gene identification by sequence homology searching in plants since a proportion of genes so identified may not be functionally active meiotic genes. Furthermore, when two or more genes originating from duplication encode functionally related proteins, mutational inactivation of one gene may be insufficient to produce a mutant phenotype. Functional analysis then has to proceed by other approaches. This general problem is not confined to *Arabidopsis*; another prominent example is the evidence for ancient genome doubling in the common ancestor of modern grasses as well as more recent doubling in the maize lineage, even though it exhibits diploid-like meiotic behavior. Neither is this issue confined to plants since there is evidence of ancestral genome duplication in other organisms/organismal groups, including yeast and mouse, but the indications are that this has been and continues to be more prevalent in plants.

According to a recent Royal Society review, "Reaping the Benefits: Science and the Sustainable Benefits of Global Agriculture," the global demand for food is likely to double by 2050. This will require breeding of new crops with improved yields and other desirable traits. The manipulation of genetic recombination in crops is likely to make an important contribution towards achieving this goal. The advances made in our understanding of plant meiosis arising from the application of molecular techniques described in this volume, in conjunction with cytological studies, should ensure that we are well placed to meet this challenge.

Birmingham, UK   **Gareth Jones**

Preface

Meiosis is one of the most critical processes in eukaryotes, required for continuation of species and generation of new variation. In plants, meiotic recombination is by far the most important source of genetic variation. For the past several years, we have been witnessing a revolution in our understanding of how meiosis works in plants, particularly at the molecular level. These insights have been made possible by advances in methods for molecular cytogenetics and chromosome analysis. This new set of tools helps us understand the organization and behavior of the genetic material in a wide range of both model and crop species. In this volume, we have assembled an extensive list of protocols developed and used in a number of laboratories at the cutting edge of meiosis and chromosome research. We are highly indebted to all contributors for the work they put into compiling the protocols and their willingness to share them with the scientific community. We hope that this book will be a useful addition to the library of both established and newly set up laboratories.

Ithaca, NY, USA
Versailles, France
Birmingham, UK

Wojciech P. Pawlowski
Mathilde Grelon
Susan Armstrong

Contents

Contributors

Patrice S. Albert • *Division of Biological Sciences, University of Missouri–Columbia, Columbia, MO, USA*

Lorinda K. Anderson • *Department of Biology, Colorado State University, Fort Collins, CO, USA*

Susan Armstrong • *School of Biosciences, The University of Birmingham, Birmingham, UK*

Philippa Barrell • *New Zealand Institute for Plant and Food Research, Christchurch, New Zealand*

Hank W. Bass • *Department of Biological Science, The Florida State University, Tallahassee, FL, USA; Institute of Molecular Biophysics, The Florida State University, Tallahassee, FL, USA*

James A. Birchler • *Division of Biological Sciences, University of Missouri–Columbia, Columbia, MO, USA*

Liudmila A. Chelysheva • *Institut National de la Recherche Agronomique (INRA), Centre de Versailles-Grignon, Versailles Cedex, France*

Changbin Chen • *Department of Horticulture, University of Minnesota, St. Paul, MN, USA*

Zhukuan Cheng • *State Key Laboratory of Plant Genomics and Center for Plant Gene Research, Institute of Genetics and Developmental Biology, Chinese Academy of Sciences, Beijing, China*

Wayne Crismani • *Institut National de la Recherche Agronomique (INRA), Centre de Versailles-Grignon, Institut Jean-Pierre Bourgin UMR1318 INRAAgroParisTech, Versailles Cedex, France*

Tatiana V. Danilova • *Division of Biological Sciences, University of Missouri–Columbia, Columbia, MO, USA*

R. Kelly Dawe • *Department of Plant Biology, University of Georgia, Athens, GA, USA; Department of Genetics, University of Georgia, Athens, GA, USA*

Jan Drouaud • *Institut National de la Recherche Agronomique (INRA), Centre de Versailles-Grignon, Institut Jean-Pierre Bourgin UMR1318 INRAAgroParisTech, Versailles Cedex, France*

F. Chris H. Franklin • *School of Biosciences, The University of Birmingham, Birmingham, UK*

Zhi Gao • *Division of Biological Sciences, University of Missouri–Columbia, Columbia, MO, USA*

Laurène Giraut • *Institut National de la Recherche Agronomique (INRA), Centre de Versailles-Grignon, Institut Jean-Pierre Bourgin UMR1318 INRAAgroParisTech, Versailles Cedex, France*

Laurie Grandont • *Institut National de la Recherche Agronomique (INRA), Centre de Versailles-Grignon, Versailles Cedex, France*

Mathilde Grelon • *Institut National de la Recherche Agronomique (INRA), Centre de Versailles-Grignon, Versailles Cedex, France*

UELI GROSSNIKLAUS • *Institute of Plant Biology and Zürich-Basel Plant Science Center, University of Zürich, Zürich, Switzerland*

FANGPU HAN • *Division of Biological Sciences, University of Missouri–Columbia, Columbia, MO, USA*

YAN HE • *Cornell University, Ithaca, NY, USA*

JAMES D. HIGGINS • *School of Biosciences, The University of Birmingham, Birmingham, UK*

ELIZABETH S. HOWE • *Department of Biological Science, The Florida State University, Tallahassee, FL, USA*

ELAINE C. HOWELL • *School of Biosciences, The University of Birmingham, Birmingham, UK*

GARETH JONES • *School of Biosciences, The University of Birmingham, Birmingham, UK*

HOSSEIN KHADEMIAN • *Institut National de la Recherche Agronomique (INRA), Centre de Versailles-Grignon, Institut Jean-Pierre Bourgin UMR1318 INRAAgroParisTech, Versailles Cedex, France*

JONATHAN C. LAMB • *Division of Biological Sciences, University of Missouri–Columbia, Columbia, MO, USA*

MARTIN A. LYSAK • *Laboratory of Plant Cytogenomics, Central European Institute of Technology (CEITEC), Masaryk University, Brno, Czech Republic*

CHRISTOPHER A. MAKAROFF • *Department of Chemistry and Biochemistry, Miami University, Oxford, OH, USA*

TEREZIE MANDÁKOVÁ • *Laboratory of Plant Cytogenomics, Central European Institute of Technology (CEITEC), Masaryk University, Brno, Czech Republic*

KARL MECHTLER • *Institute of Molecular Biotechnology (IMBA) and Institute of Molecular Pathology (IMP), Vienna, Austria*

RAPHAËL MERCIER • *Institut National de la Recherche Agronomique (INRA), Centre de Versailles-Grignon, Institut Jean-Pierre Bourgin UMR1318 INRAAgroParisTech, Versailles Cedex, France*

CHRISTINE MÉZARD • *Institut National de la Recherche Agronomique (INRA), Centre de Versailles-Grignon, Institut Jean-Pierre Bourgin UMR1318 INRAAgroParisTech, Versailles Cedex, France*

SHAUN P. MURPHY • *Institute of Molecular Biophysics, The Florida State University, Tallahassee, FL, USA*

KIM OSMAN • *School of Biosciences, The University of Birmingham, Birmingham, UK*

WOJCIECH P. PAWLOWSKI • *Department of Plant Breeding and Genetics, Cornell University, Ithaca, NY, USA*

ERNEST F. RETZEL • *National Center for Genome Research, Santa Fe, NM, USA*

ELISABETH ROITINGER • *Institute of Molecular Pathology (IMP), Vienna, Austria*

MOIRA J. SHEEHAN • *Cornell University, Ithaca, NY, USA*

GAGANPREET SIDHU • *Cornell University, Ithaca, NY, USA*

STEPHEN M. STACK • *Department of Biology, Colorado State University, Fort Collins, CO, USA*

LJUDMILLA TIMOFEJEVA • *Department of Gene Technology, Tallinn University of Technology, Tallinn, Estonia*

CHUNG-JU RACHEL WANG • *Institute of Plant and Microbial Biology, Academia Sinica, Taipei, Taiwan*

JIANHUA YANG • *School of Biosciences, The University of Birmingham, Birmingham, UK*
XIAOHUI YANG • *Department of Chemistry and Biochemistry, Miami University, Oxford, OH, USA*
LI YUAN • *Department of Chemistry and Biochemistry, Miami University, Oxford, OH, USA*

Part I

Cytological Techniques for Light Microscopy

Spreading and Fluorescence In Situ Hybridization of Male and Female Meiocyte Chromosomes from *Arabidopsis thaliana* for Cytogenetical Analysis

Susan Armstrong

Abstract

Advances in molecular biology and in the genetics of *Arabidopsis thaliana* have led to it becoming an important model for the analysis of meiosis in plants. Cytogenetic investigations are pivotal to meiotic studies and a number of technological improvements for *Arabidopsis* cytology have provided a range of tools to investigate chromosome behavior during meiosis. This chapter contains a detailed description of cytological techniques currently used in our laboratory for the basic preparation of meiotic chromosomes for investigation of the female and male meiotic pathway and fluorescence in situ hybridization (FISH) analysis for the frequency and distribution of crossovers (chiasmata) at metaphase I.

Keywords Cytological techniques, *Arabidopsis thaliana*, Female meiosis, Male meiosis, Chromosomes, FISH

1 Introduction

The developments in cytological techniques described in this chapter combined with the extensive range of genomic resources that are available have established *Arabidopsis thaliana* as an excellent system for the analysis of meiosis (1, 2).

Female and male meiosis in *Arabidopsis*, similar to many plant species, occurs asynchronously within the developing flower bud. The flower has a simple structure, typical of the *Brassicaceae*, having four free sepals and four petals, four long medial stamens and two shorter lateral stamens. The gynoecium is superior and composed of two carpels with locules separated by a false septum (3, 4). Inflorescences in all plant species require a methodological approach to locate and identify meiotic stages this is particularly true of *Arabidopsis* where we are dealing with small bud sizes. We have previously linked meiotic stages with floral development (3) and reproduce that table with modifications (Table 1).

Wojciech P. Pawlowski et al. (eds.), *Plant Meiosis: Methods and Protocols*, Methods in Molecular Biology, vol. 990, DOI 10.1007/978-1-62703-333-6_1, © Springer Science+Business Media New York 2013

Table 1
Summary of the relationship between floral development and cytological landmarks

Floral stage (as described by Smyth et al. (3))	Bud size (mm)	Gynoecium length during meiosis	Cytology	Meiotic stage
(8) Stamens already developed into anthers and filaments (stage 7). Primordia of petals visible. Sepals cover tip of bud	<0.3		Anthers differentiated into five tissue layers; pollen mother cells are surrounded by the tapetum, the middle cell layer, the endothecium, and the epidermis	Meiotic inter-phase in pollen mother cells
(9) Petal primordia stalked at base	0.3–0.4		Tapetal cells become binucleate at zygo-tene/early pachytene stages	Meiosis in pollen mother cells
(10) Petals level with two lateral, shorter stamens. Stamens green	<0.5	0.3–0.4	Gynoecium cylinder deeply slotted	Microspores. Meiotic inter-phase in embryo mother cells
		0.5	Closure of cylinder, appearance of stigmatic papillae	Meiosis in embryo mother cells
(11) Anthers yellow	<0.7	0.6–0.8	Stigmatic papillae develop	Meiosis in embryo mother cells

Modified from Ref. 4

This chapter describes our current protocols, including trouble-shooting for the cytological analysis of *Arabidopsis* chromosomes. The protocol described here includes the basic preparation of mei-otic chromosomes for investigation of the female and male meiotic pathway (5, 6) and using fluorescence in situ hybridization (FISH) to identify the individual *Arabidopsis* chromosome pairs (7, 8).

2   Materials

2.1   Plants

Sow seeds of *Arabidopsis* in 6 cm diameter pots in soil-based com-post. The plants should be grown in dedicated growth chambers maintained at 18°C with a 16 h light cycle. In our conditions, the plants arrive at flowering stages at 5–6 weeks.

2.2 Fixation and Preparation of Slides for Basic Cytology and Fluorescence In Situ Hybridization[1]

1. Fixative: 3 parts of absolute ethanol:1 part of glacial acetic acid, keep on ice. Prepare fresh fixative as required and discard at the end of the day.

2. 0.01 M Citrate Buffer: prepare a working solution of the buffer (pH 4.5) by using 4.45 ml of 0.1 M sodium citrate, 5.55 ml of 0.1 M citric acid, made up to 100 ml with sterile deionized water.

3. Stock digestion medium: dissolve 1% cellulase (Sigma Aldrich, St. Louis, MO, USA) and 1% pectolyase (Sigma Aldrich, St. Louis, MO, USA) in a working solution of 0.01 M citrate buffer, pH 4.5. Dispense in aliquots and store at –20°C.

4. Digestion medium: mix 333 µl of the stock digestion medium with 667 µl of 0.01 M citrate buffer, pH 4.5.

5. 60% acetic acid: dilute glacial acetic acid with sterile deionized water.

6. Counterstaining solution: mix 10 µl/ml of 1 mg/ml 4,6-diaminido-2-phenlyinidole (DAPI) in an antifade mounting medium, Vectashield (Vector Laboratories, Burlingame, CA, USA). DAPI should be stored as a stock solution at 1 mg/ml in sterile deionized water. Dispense in aliquots and store at –20°C.

2.3 FISH Analysis

Any single copy DNA clone over the size of around 2 kb can be used for FISH analysis. We frequently use two DNA repetitive clones, 45S rDNA and 5S rDNA (8), which, when used simultaneously, enable us to identify each bivalent unequivocally. This method can be used, for example, for the estimation of crossover frequency and distribution in *Arabidopsis* by counting chiasmata at metaphase I.

1. Pretreatment washing solution: 2× SSC buffer (0.3 M NaCl, 0.03 M sodium citrate, pH 7.0). Prepare a stock solution of 20× SSC and store at room temperature.

2. Digestion medium: 0.01 g of pepsin (MP Biomedicals, Irvine, CA, USA) in 100 ml of 0.01 M HCl. Prepare freshly and pre-heat for a few minutes at 37°C before use.

3. Paraformaldehyde fixative: weigh out 4 g paraformaldehyde (electron microscopy grade) in the fume hood. Dissolve in 100 ml of sterile deionized water and four drops of 1 M NaOH, pre-warmed to 60°C in the microwave. Stir the mixture on a magnetic stirrer for 1 h, or until it dissolves, before filtering through Whatman paper. Adjust the pH to 8.0. The fixative can be stored for up to 1 week at 4°C.

---

[1] Adapted with kind permission from Armstrong, S.J., Sanchez-Moran, E., and Franklin, F.C.H. (2009) Cytological analysis of *Arabidopsis thaliana* meiotic chromosomes. *Methods in Molecular Biology 558: Meiosis Cytological Methods*, 131–145.

4. Alcohol series made up with absolute ethanol and sterile deionized water: 70%, 90%, and 100% ethanol, used for slide dehydration.

5. Probes: 5S ribosomal DNA. Plasmid pCT4.2 containing the 5S rDNA gene from *Arabidopsis thaliana* (8, 9).

6. 45S ribosomal DNA. Clone pTa71 (10) containing a 9 kb *Eco*RI fragment from *Triticum aestivum* consisting of the 18S–25S rRNA genes and their spacer regions.

7. Probe labelling: use the nick translation labelling kit (Roche Applied Science, Penzberg, Germany) following the manufacturer's instructions. Use either Biotin-16-dUTP or digoxigenin-11-dUTP (Roche Applied Science, Penzberg, Germany) as nucleotide conjugates for DNA labelling.

8. Hybridization mix: mix 1 g dextran sulfate (use high MW 500,000), 5 ml deionized formamide, (see Note 1), and 1 ml 20× SSC. Make up to 7 ml with sterile deionized water. Dissolve the dextran sulfate at 65°C, cool and pH to 7.0. Aliquot the solution into 1 ml Eppendorf tubes and store at –20°C.

9. Prepare 20 µl of probe mixture per slide: combine 14 µl of hybridization mix, 0.5–2 µl of labelled probe, and, if necessary, add sterile deionized water to 20 µl. The final hybridization mix should consist of 50% deionized formamide, 2× SSC, and 10% dextran sulfate pH 7.0.

10. Vulcanizing rubber solution: (e.g., as found in bicycle wheel repair kits).

11. Prepare post-hybridization washes: three Coplin jars containing 50% formamide, 2× SSC pH 7.0 (150 ml of deionized formamide, 30 ml of 20× SSC, 120 ml of sterile deionized water), and one Coplin jar of 2× SSC along with 1,000 ml of 4× SSC wash (200 ml of 20× SSC, 800 ml sterile deionized water, and 0.5 ml Tween 20).

12. For detection of digoxigenin probes, prepare antibodies as either anti-digoxigenin-FITC or anti-digoxigenin rhodamine at the concentration of 5 ng/µl in digoxigenin blocking solution. Prepare shortly before use. The blocking solution is made from 100 ml of 4× SSC, 0.05% Tween 20, and 0.5% Roche digoxigenin blocking reagent (Roche Applied Science, Penzberg, Germany). Centrifuge at 19,000 ×*g* for 5 min, and store the supernatant in 1 ml aliquots at –20°C.

13. For biotin labelled probes, use Streptavidin-Texas Red/Cy3/FITC (Roche Applied Science, Penzberg, Germany) in biotin blocking solution. The biotin blocking solution is made from 100 ml of 4× SSC, 0.05% Tween 20, and 5 g of dried skimmed milk. Centrifuge at 19,000 ×*g* for 5 min and store the supernatant in 1 ml aliquots at –20°C.

14. Counterstaining solution: see Subheading 2.2, item 6.

3 Methods

3.1 Fixation and Preparation of Slides for Basic Cytology and FISH Analysis

1. Cut terminal inflorescences with two or three open flowers only and immediately place them in freshly made ice-cold fixative. Leave them on the bench at room temperature and replace the fix after 2–3 h. The fixed material is suitable for digestion 24 h after fixation, and can be stored at –20°C for up to 6 months.

2. Place single fixed inflorescences in watch glasses (preferably colored black) with fresh fixative. Using a mounted needle and fine forceps (e.g., watchmaker's forceps), divide up the inflorescence into individual buds, and order them according to size. For male meiosis, retain buds in the range of 0.3–0.4 mm. Buds with yellow anthers are unsuitable (Table 1).

3. For female meiosis, dissect gynoecia in the range of 0.3–0.9 mm from buds of sizes 0.5–0.7 mm (Table 1) and place them in a watch glass.

4. Replace the fixative with the working citrate buffer (2 × 5 min) and then with the working solution of the enzymes. Incubate in a humidified atmosphere, e.g., a sandwich box containing damp tissues, at 37°C for 75–90 min for the buds (shorter times preserve the organization of the pollen mother cells in early meiosis) and for female meiosis, for 30–60 min (again shorter duration in enzyme mixture preserves the organization of the tissues around the embryo mother cell).

5. Remove the enzyme solution and replace it with the working citrate buffer at 4°C in order to stop the reaction.

6. Place a single bud or gynoecium onto a slide, with a minimum of buffer. Macerate the bud quickly with a mounted needle, ensuring that the material does not dry out.

7. Add 10 µl of 60% acetic acid, mix with the material on the slide, and place on a hot block at 45°C for up to 30 s, (see Note 2). Circle the region of the slide containing the material with a diamond pen.

8. Place the slide on the bench and add 100 µl of ice-cold fixative as a ring around the material. Dry the preparation. We find using a commercial hair drier suitable for this purpose.

9. The slides are now ready for basic cytology after mounting in 7 µl of DAPI in Vectashield, and view with a fluorescence microscope and the images captured with an image analysis system (Fig. 1).

3.2 FISH (See Notes 3–5)

1. Wash the chromosome preparations in 2× SSC at room temperature for 10 min.

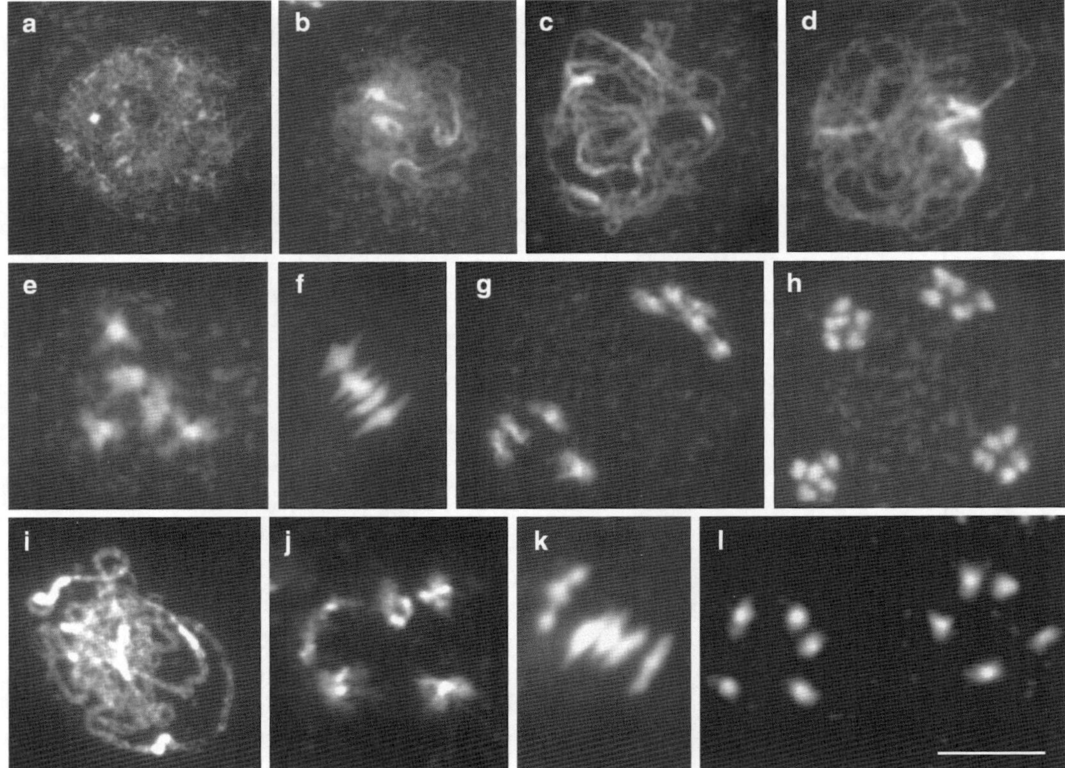

Fig. 1 Male and female meioses in wild-type *Arabidopsis* prepared by the spreading technique and stained with DAPI. (**a–h**) Male meiosis in wild-type pollen mother cells (PMCs). Leptotene showing unsynapsed chromosome axes. (**a**). Late zygotene, showing mostly synapsed axes and a few unsynapsed regions. Note the typical distribution of mitochondria in zygotene, which are located around one of the poles only. (**b**). Pachytene, showing full synapsis. (**c**). Early diplotene, showing some homologue separation. Note distribution of the mitochondria, randomly located throughout the cytoplasm. (**d**). Diakinesis showing five moderately condensed, unaligned bivalents. (**e**). Metaphase I, showing five condensed and aligned bivalents. The number of crossovers (COs) in this cell having five ring bivalents is 10. (**f**). Metaphase II showing two groups of five chromosomes. (**g**). Tetrad stage showing 4 haploid nuclei at the end of meiosis. (**i–l**) female meiosis in wild-type embryo-mother cells (EMCs). (**h**). Pachytene, showing full synapsis. (**i**). Diakinesis showing five moderately condensed, unaligned bivalents. (**j**). Metaphase I, showing five condensed and aligned bivalents. The number of COs in this cell, having three rod bivalents and two ring bivalents is 7. (**k**). Metaphase II showing two groups of five chromosomes. Bar = 10 μm

2. Digest the preparations with pepsin at 37°C for 45–90 s. Do not over-digest or you will lose the material from the slide.

3. Wash in 2× SSC at room temperature twice for 5 min.

4. Fix the material in paraformaldehyde in the fume hood for 10 min.

5. Wash twice in sterile deionized water.

6. Dehydrate the preparations by passing them through an alcohol series, 70%, 90% and 100%, 2 min each wash.

7. Drain the slides. Check if the material is still on the slides using a phase-contrast microscope.

8. To denature the probe and chromosomes, place 20 µl of the probe mixture on the slide and seal with a coverslip (22 × 22 mm) and vulcanizing rubber solution. Heat the slides on a hotplate at 75°C for 4 min.

9. For hybridization, incubate the slides in a humidified atmosphere at 37°C overnight.

10. To conduct post-hybridization washes, remove the rubber solution and coverslips using fine forceps. Wash the slides three times, 5 min for each wash, in 50% formamide, 2× SSC at 45°C, then once in 2× SSC at 45°C for 5 min, and afterwards once in 4× SSC, 0.05% Tween 20 at 45°C, 5 min and once in 4× SSC, 0.05% Tween 20 at room temperature for 5 min.

11. Labelling detection: add fluorescently labelled anti-digoxigenin or anti-streptavidin antibodies to the slides (100 µl per slide), cover the slides with a rectangle of parafilm to fit the slide, and incubate them in a incubated in a humidified atmosphere at 37°C in the dark for 30 min. Afterwards, remove the parafilm and wash the preparations in the washing solution three times, 5 min each wash.

12. Counterstain slides with 7 µl of DAPI staining solution.

13. View the FISH preparations with a fluorescence microscope equipped with filters for DAPI, TRITC, and FITC and an image capture and analysis system.

14. To estimate recombination rates in *Arabidopsis* by counting chiasmata at metaphase I, see Fig. 2.

4   Notes

1. We deionize formamide by mixing 200 ml of formamide with 10 g of mixed resin beads (PlusOne Amberlite IRN-150 L, GE Healthcare, Piscathaway, NJ, USA) on a stirrer for at least 1 h. Filter and store at 4°C.

2. When making slides of fixed material, we use 60% acetic acid to dissociate the cells. Too long incubation on the hot block may cause damage to the cells and loss of material.

3. We usually take between five and eight slides through the pre-treatments using an appropriate Coplin jar. The preparations are generally left on the bench overnight. Slides that have been stored at room temperature for more than 4 weeks give less satisfactory labelling.

4. For experiments that require slides with different meiotic stages, prescreen the slides with a phase contrast microscope.

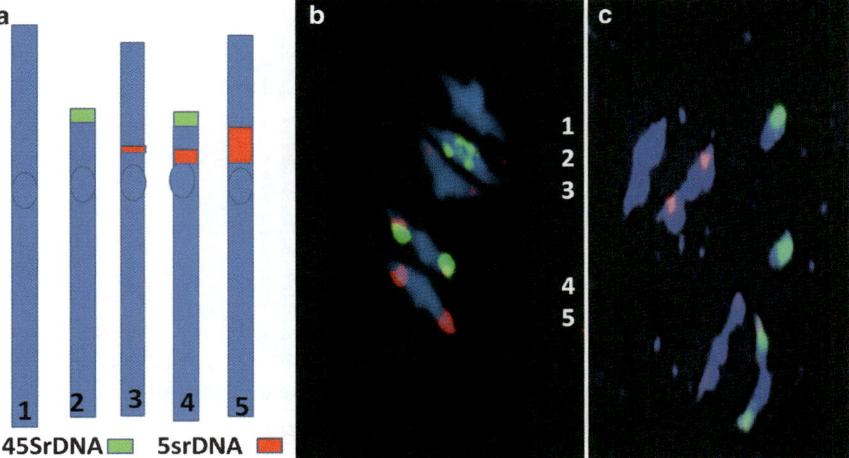

Fig. 2 Using metaphase I bivalents to estimate crossover distribution and frequency. (**a**) Haploid idiogram of *Arabidopsis thaliana* (COL-0) showing the location of 5S rDNA (*red*) and 45S rDNA (*green*). (**b**, **c**) Using these two repetitive probes enables unequivocal identification of each chromosome pair. (**b**) Wild type metaphase I, with each bivalent chromosome pair numbered: chromosome 1 with 3 crossovers, (distal and distal + interstitial) forming a ring, chromosome 2 with two distal crossovers forming a ring, chromosome 3 with three crossovers, (distal and interstitial + distal) forming a ring, and chromosomes 4 and 5 both forming rod bivalents with a single crossover. (**c**) Meiotic mutant *Atmlh3* with low crossover frequency compared to wild type: chromosome 1 with two distal crossovers forming a ring, chromosome 2 with 2 univalents, chromosomes 3, 4, 5 all rod bivalents with single crossovers

The slides are prescreened using a phase contrast microscope. Less experienced cytogeneticists may prefer to use DAPI for prescreening; these slides need to be destained before the FISH protocol as follows: wash 10 min in 2× SSC to remove coverslips, then wash slides through an alcohol series of 70%, 85% and 100% ethanol, 2 min in each solution. The slides are ready for the FISH protocol.

5. For estimation of recombination frequency by counting chiasmata, we use slides containing metaphase I meiocytes. Female metaphase I cells tend to be difficult to find and the slides require prescreening, we often only find three to five analyzable metaphase I cells per slide. We do not analyze diakinesis chromosomes because of the problem of distinguishing a crossover from a twisted chromosome.

Acknowledgements

The work in our laboratory is supported by the Biotechnology and Biological Sciences Research Council (BBSRC) and the European Commission, FP7 grant 222883. Technical support has been provided by Steve Price.

References

1. Arabidopsis Genome Initiative (2000) Analysis of the genome sequence of the flowering plant *Arabidopsis thaliana*. Nature 408:796–815

2. Jones GH, Armstrong SJ, Caryl AP, Franklin FCH (2003) Meiotic chromosome synapsis and recombination in *Arabidopsis thaliana*; an integration of cytological and molecular approaches. Chromosome Res 11:205–215

3. Smyth DR, Bowman JL, Meyerowitz EM (1990) Early flower development in *Arabidopsis*. Plant Cell 2:755–767

4. Armstrong SJ, Jones GH (2003) Meiotic cytology and chromosome behaviour in wild-type *Arabidopsis thaliana*. J Exp Bot 54:1–10

5. Ross KJ, Fransz P, Jones GH (1996) A light microscopic atlas of meiosis in *Arabidopsis thaliana*. Chromosome Res 4:507–516

6. Armstrong SJ, Jones GH (2001) Female meiosis in *Arabidopsis thaliana* and in two meiotic mutants. Sex Plant Reprod 13:177–183

7. Fransz P, Armstrong SJ, Alonso-Blanco C, Fischer TC, Torrez-Ruiz RA, Jones GH (1998) Cytogenetics for the model system *Arabidopsis thaliana*. Plant J 13:867–876

8. Sanchez-Moran E, Armstrong SJ, Santos JL, Franklin FCH, Jones GH (2002) Variation in chiasma frequency among eight accessions of *Arabidopsis thaliana*. Genetics 162:1415–1422

9. Campbell BR, Song Y, Posch TE, Cullis CA, Town CD (1992) Sequence and organisation of 5S ribosomal RNA-encoding genes of *Arabidopsis thaliana*. Gene 112:225–228

10. Gerlach WL, Bedrock JR (1979) Cloning and characterization of ribosomal RNA genes from wheat and barley. Nucleic Acids Res 7:1869–1885

Chapter 2

Analysis of Plant Meiotic Chromosomes by Chromosome Painting

Martin A. Lysak and Terezie Mandáková

Abstract

Chromosome painting (CP) refers to visualization of large chromosome regions, entire chromosome arms, or entire chromosomes via fluorescence in situ hybridization (FISH). For CP in plants, contigs of chromosome-specific bacterial artificial chromosomes (BAC) from the target species or from a closely related species (comparative chromosome painting, CCP) are typically applied as painting probes. Extended pachytene chromosomes provide the highest resolution of CP in plants. CP enables identification and tracing of particular chromosome regions and/or entire chromosomes throughout all meiotic stages as well as corresponding chromosome territories in premeiotic interphase nuclei. Meiotic pairing and structural chromosome rearrangements (typically inversions and translocations) can be identified by CP. Here, we describe step-by-step protocols of CP and CCP in plant species including chromosome preparation, BAC DNA labeling, and multicolor FISH (**Fig. 1**).

Keywords Chromosome painting, Fluorescence in situ hybridization, BAC FISH, Pachytene chromosomes, DNA labeling

1 Introduction

The chromosome painting (CP) technique in human and animal cytogenetics serves in situ identification of whole chromosomes and specific chromosome regions using flow sorted, Degenerate Oligonucleotide Primed PCR (DOP-PCR)-amplified, and fluorescently labeled chromosomes or chromosome segments. Due to abundant and diverse DNA repeats homogeneously distributed across chromosome complements (1), flow sorted or microdissected chromosomes are not suitable for preparation of chromosome-specific painting probes in plants. Instead, chromosome-specific high-capacity DNA vectors have become widely utilized in plant cytogenetics since the mid-1990. Specifically, individual bacterial artificial chromosome (BAC) clones and BAC contigs (continuous sets of BAC clones) are most frequently used

Wojciech P. Pawlowski et al. (eds.), *Plant Meiosis: Methods and Protocols*, Methods in Molecular Biology, vol. 990, DOI 10.1007/978-1-62703-333-6_2, © Springer Science+Business Media New York 2013

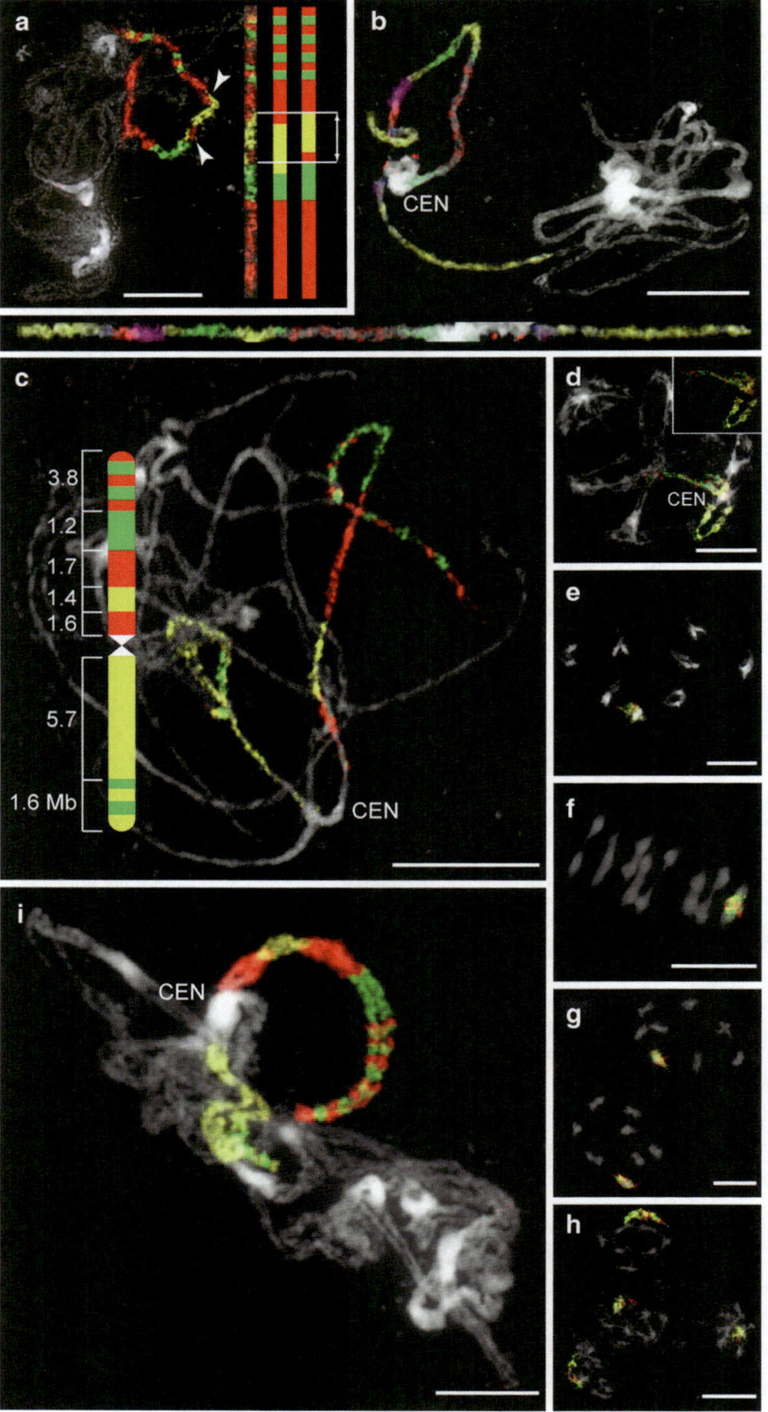

Fig. 1 Application of comparative multicolor chromosome painting for meiotic studies. (**a**) Chromosome painting in Shahdara × Columbia (Sha × Col) hybrid of *Arabidopsis thaliana* ($n = 5$, At1–At5). A 2.2 Mb paracentric inversion on the top arm of chromosome At3 is specific for Sha and absent in Col. Paired Sha and Col

as chromosome-specific probes. Fluorescence in situ hybridization (FISH) of single or several BAC clones is referred to as BAC FISH, whereas BAC painting or chromosome painting applies to in situ hybridization of BAC contigs covering larger chromosome regions (e.g., chromosome arms) or whole chromosomes (Fig. 1). Chromosome-specific BAC contigs are used as painting probes either in the same species (2–5) or in species with sufficient chromosome homeology (cross-species or comparative chromosome painting, CCP) (3, 6–8). Recently, reciprocal BAC painting and multi-species BAC painting (using painting probes of two or more species to chromosomes of another species) on pachytene chromosomes has been established (9).

Although CP enables identification of chromosome regions and whole chromosomes during all (pre)meiotic stages (2), pachytene bivalents and multivalents usually offer the highest resolution. Here, we provide a protocol to paint meiotic and mitotic chromosomes of plants using chromosome-specific BAC clones and BAC contigs. These procedures are essentially based on our long-term experience with CP and CCP in crucifer species (*Brassicaceae*) (2, 3, 6, 8, 9).

2   Materials

Prepare and store all reagents at room temperature (unless indicated otherwise).

2.1   Collection of Floral Material

1. Freshly prepared Carnoy's I fixative: 3 parts ethanol, 1 part glacial acetic acid; or Carnoy's II fixative: 6 parts ethanol, 3 parts chloroform, 1 part glacial acetic acid (see Note 1).

Fig. 1 (continued) homologues were identified by differentially labeled *A. thaliana* BAC contigs (inversion marked by *arrowheads*).(**b–i**) Comparative chromosome painting in *Brassicaceae* species using chromosome-specific *A. thaliana* BAC contigs. (**b**) Pachytene chromosome Sn4 of *Stenopetalum nutans* (n=4, Sn1–Sn4) painted by 14 differentially labeled *A. thaliana* BAC contigs. Image of the same chromosome digitally straightened using the "straighten-curved-objects" plugin in the Image J software (12). (**c–h**) CCP of chromosome Ca1 in diploid *Cardamine amara* (n=8, Ca1–Ca8) at pachytene (**c**), diplotene (**d**), diakinesis (**e**), metaphase I (**f**), telophase I (**g**), and anaphase/telophase II (**h**). (**i**) CCP of a pachytene tetravalent of Ca1 in autotetraploid *C. amara* (n=16, Ca1–Ca16). Chromosome-specific *Arabidopsis* BAC contigs were labeled by biotin-dUTP (*red*) and digoxigenin-dUTP (*green*), and immuno-detected using antibodies coupled to Texas Red and Alexa Fluor 488, respectively. *Yellow* signals correspond to Cy3-dUTP-labeled contigs. In addition, BAC clones hybridized to chromosome Sn4 in (**b**) were also labeled by DEAC-dUTP (*blue*) and DNP-dUTP (Cy5, *magenta*). CENs refer to centromeres (not labeled by BAC clones). Size of BAC contigs in Mb according to http://www.arabidopsis.org. All meiotic chromosomes were isolated from anthers and counterstained with DAPI. *Bars* = 10 μm

2. 70% ethanol.

3. Forceps.

4. Glass vials or microcentrifuge tubes.

2.2 Chromosome Preparation

1. 10× citrate buffer: 40 mL of 100 mM citric acid and 60 mL of 100 mM trisodium citrate, adjust pH to 4.8; store at 4°C.

2. Pectolytic enzyme mixture in 1× citrate buffer: 0.3% pectolyase, 0.3% cellulase, 0.3% cytohelicase (Sigma Aldrich, St. Louis, MO, USA) prepared from frozen 1% stock solutions in 1× citrate buffer (see Note 2).

3. 60% glacial acetic acid in distilled water (see Note 3).

4. Freshly prepared ice-cold Carnoy's I fixative.

5. 4% freshly prepared formaldehyde in distilled water.

6. "Assistant" staining blocks with glass cover (Karl Hecht Assistant, Sondheim/Rhön, Germany).

7. Humid box.

8. Incubator at 37°C.

9. Stereomicroscope, microscope with phase contrast.

10. SuperFrost microscope slides (Fisher Scientific, Suwanee, GA, USA).

11. Dissection needles.

12. Fine forceps.

13. Glass Pasteur pipette.

14. Heating block.

15. Hair dryer.

2.3 Chromosome Preparation Pretreatment

1. 20× Saline Sodium Citrate (SSC): 3 M sodium chloride, 300 mM trisodium citrate, pH 7.0.

2. RNase: 100 μg/mL DNase-free ribonuclease A (AppliChem, St. Louis, MO, USA). Make stock of 1 mg/mL in distilled water. Store aliquots at −20°C.

3. Pepsin from porcine gastric mucosa (Sigma Aldrich, St. Louis, MO, USA) 0.1 mg/mL in 10 mM HCl. Prepared from 100 mg/mL stock in 10 mM HCl. Store aliquots at −20°C.

4. 4% freshly prepared formaldehyde in 2× SSC.

5. 70%, 80%, and 96% ethanol.

6. 4′, 6-diamidino-2-phenylindole (DAPI; Sigma Aldrich, St. Louis, MO, USA): 2 μg/mL in Vectashield antifade (Vector Laboratories, Burlingame, CA, USA). Store at 4°C.

7. Coverslips (24×24 mm and 24×50 mm).

8. Coplin or Hellendahl jar (see Note 4).

9. Humid box.

10. Incubator and water bath at 37°C.

11. Plastic tube rack.

2.4 Probe Labeling

1. 10× NT buffer: 500 mM Tris–HCl pH 7.5, 50 mM $MgCl_2$, 0.05% bovine serum albumin.

2. Nucleotide mixture: 2 mM dATP, dCTP, dGTP, and 400 mM dTTP (Roche Applied Science, Indianapolis, IN, USA).

3. 1 mM x-dUTP (x stands for biotin, digoxigenin, Cy3 or other hapten/fluorochrome; see Note 5).

4. 0.1 M β-mercaptoethanol.

5. DNase I (Roche Applied Science, Indianapolis, IN, USA). Use a 1:250 dilution of a 1 mg/mL DNase stock in 0.15 M NaCl in 50% glycerol.

6. DNA polymerase I (10 U/µL; Fermentas, Glen Burnie, MA, USA).

7. 0.5 M EDTA, pH 8.0.

8. 100 bp DNA ladder.

9. 0.5 mL microcentrifuge tubes.

10. Thermocycler for 15°C and 60°C.

11. Electrophoresis system.

12. 1% agarose gel.

2.5 Probe Preparation and In Situ Hybridization

1. 3 M sodium acetate, pH 5.2.

2. 70% ethanol, 96% ice-cold ethanol.

3. Hybridization buffer: 50% formamide, 10% dextran sulfate in 2× SSC.

4. 2 mL microcentrifuge tubes.

5. Refrigerated centrifuge.

6. Vacuum desiccator or SpeedVac.

7. Thermomixer (a heating block with exact temperature control) at 80°C.

8. 24 × 24 mm, 22 × 22 mm, or 24 × 32 mm coverslips.

9. Rubber cement.

10. Humid box.

11. Incubator at 37°C.

2.6 Fluorescence Detection of Hybridized Probes

1. 2× SSC.

2. 50% or 20% deionized formamide in 2× SSC, pH 7.0 (see Note 6).

3. 4T buffer: 4× SSC pH 7.0, 0.05% Tween-20.

4. Blocking solution: 5% bovine serum albumin, 0.2% Tween-20 in 4× SSC.

5. TNT buffer: 100 mM Tris–HCl pH 7.5, 150 mM NaCl, 0.05% Tween-20.

6. TNB buffer: 100 mM Tris–HCl pH 7.5, 150 mM NaCl, 0.5% blocking reagent (Roche Applied Science, Indianapolis, IN, USA).

7. Antibodies: avidin–Texas Red (Vector Laboratories, Burlingame, CA, USA), goat anti-avidin–biotin (Vector Laboratories, Burlingame, CA, USA), mouse anti-digoxigenin (Jackson ImmunoResearch Laboratories, West Grove, PA, USA), goat anti-mouse–Alexa Fluor 488 (Invitrogen, Carlsbad, CA, USA).

8. DAPI (2 μg/mL) in Vectashield antifade.

9. 70%, 80% and 96% ethanol.

10. Coplin or Hellendahl jars (see Note 2).

11. Water bath shaker at 42°C.

12. Coverslips (24 × 32 mm and 24 × 50 mm).

13. Humid box.

14. Incubator at 37°C.

15. Epifluorescence microscope equipped with optical filters for DAPI and other fluorochromes, a digital charge-coupled device (CCD) camera, and image acquisition software.

3   Methods

Carry out all procedures at room temperature unless otherwise specified. To minimize fluorochrome bleaching, avoid overexposure to light during procedures 3.4–3.6.

3.1 Collection of Floral Material

1. Fix entire inflorescences, individual flower buds, or anthers in Carnoy's fixative at room temperature or at 4°C overnight; change the fixative several times.

2. Exchange the fixative for 70% ethanol and store the fixed material at –20°C (see Note 7).

3.2 Chromosome Preparation

1. Rinse floral material with distilled water in a staining block or small petri dish for 10 min. Under a stereomicroscope, remove and discard unwanted parts (e.g., yellow anthers containing pollen).

2. Replace water with 1× citrate buffer and wash two times, 5 min each wash.

3. Incubate the material in ~1 mL of pectolytic enzyme mixture in a humid box at 37°C for 3 h (see Note 8).

4. Replace the enzymes with 1× citrate buffer and keep the material on ice or at 4°C until use (see Note 9).

5. Put a single flower bud/anther on a microscope slide using a Pasteur pipette, remove the excess fluid and add ~20 µL of 60% acetic acid. Disintegrate the bud by dissection needles until a fine suspension is formed.

6. Place the slide on a heating block (50°C) and spread the suspension by careful circular stirring with a needle for ~30 s (see Note 10).

7. Fix the chromosomes by pipetting 100 µL of Carnoy's I fixative around the suspension drop. Discard the fluid by tilting the slide and dry using a hair dryer.

8. Using a phase-contrast light microscope, examine the preparation for specific meiotic stages and the amount of cytoplasm.

9. Post-fix the slides in a Coplin or Hellendahl jar with 4% formaldehyde in distilled water for 10 min and leave to air-dry. Carry out this step in a fume hood.

10. Store dried slides in a dust-free box at 4°C (see Note 11).

3.3 Chromosome Preparation Pretreatment

1. Wash slides two times in 2× SSC in a Coplin jar, 5 min each wash.

2. Add 100 µL of RNase solution, cover with 24 × 50 mm coverslip and incubate the slides in a humid box at 37°C for 1 h (see Note 12).

3. Tilt slides to let the coverslip fall off and wash slides as in step 1.

4. To remove cytoplasm, treat slides with pepsin at 37°C for 5 min in a Coplin jar placed in a water bath.

5. Wash slides as in step 1.

6. Dehydrate slides in an ethanol series (70%, 80%, and 96% ethanol, 3 min each) and leave them to air-dry.

7. Apply 15 µL of Vectashield with DAPI to a slide and cover it with a 24 × 24 mm coverslip. Check the slide with a fluorescence microscope. Chromosomes should be undamaged and free of cytoplasm. When cytoplasm is persistent, remove the coverslip using running water and repeat steps 4–7.

8. Remove the coverslip using running water flow and wash slides as in step 1.

9. Post-fix slides in a Coplin or Hellendahl jar with 4% formaldehyde in 2× SSC for 10 min. Carry out this and the following step in a fume hood.

10. Wash slides as in step 1.

11. Dehydrate slides in an ethanol series (70%, 80%, and 96% ethanol, 3 min each) and leave them to air-dry.

3.4 Probe Labeling by Nick Translation (See Note 13)

1. Combine in an 0.5 mL microcentrifuge tube: 1 µg of BAC DNA in 32 or 29 µL distilled water, 5 µL of 10× NT buffer, 5 µL of nucleotide mixture, 1 µL of 1 mM comercial x-dUTP or 4 µL of 1 mM custom-made x-dUTP, 5 µL of 0.1 M β-mercaptoethanol, 1 µL of DNase I, and 1 µL of DNA polymerase I. Vortex gently and spin briefly (see Note 14).

2. Incubate at 15°C for 90 min.

3. Transfer the tube on ice and load 5 µL of the reaction volume on a 1% agarose gel along with a 100-bp DNA ladder and run the electrophoresis.

4. When the smear of labeled fragments is ~200–500 bp in size, stop the nick translation by adding 1 µL of 0.5 M EDTA and heating at 60°C for 10 min. When fragments are longer than 500 bp, extend the incubation at 15°C for further 30 min and repeat steps 3 and 4.

5. Store the probe at −20°C until use.

3.5 Probe Preparation and In Situ Hybridization

1. Pool individually labeled BAC clones by pipetting 5 µL (~100 ng of DNA) of each BAC into a 2 mL microcentrifuge tube.

2. To reduce the probe volume and remove unincorporated nucleotides, precipitate the DNA by adding 0.1 volume of 3 M sodium acetate and 2.5 volume of ice-cold 96% ethanol. Vortex and keep on ice or at −20°C for at least 30 min.

3. Centrifuge at $13,000 \times g$ at 4°C for 30 min.

4. Carefully discard the supernatant.

5. Add 500 µL of 70% ethanol and centrifuge again for 5 min.

6. Repeat step 4.

7. Dry the pellet using a desiccator or SpeedVac.

8. Resuspend the pellet in 20 µL of hybridization buffer and incubate at 37°C in a thermomixer (see Note 15).

9. Add 20 µL of probe to the chromosome preparation (see Subheading 3.3), cover with a cover slip and seal with rubber cement around the edges.

10. Denature the probe and the chromosomal DNA by placing the slide on a heating block at 80°C for 2 min.

11. Incubate the slide in a humid box at 37°C overnight (see Note 16).

3.6 Fluorescence Detection of Hybridized Probes

Signal detection and amplification is described here for hapten-labeled probes (biotin- and digoxigenin-dUTP) visualized by indirect immunofluorescence via Texas Red and Alexa Fluor 488, respectively (dinitrophenyl (DNP)-dUTP can also be used as a third hapten). This protocol can be also used with fluorochrome-labeled probes (e.g., Cy3- or DEAC-dUTP) that require only post-hybridization washing prior to microscopic evaluation (steps 1–4, and 14). All washing steps are carried out at 42°C in Coplin or Hellendahl jars placed in a water bath shaker. All incubation steps are carried out at 37°C on microscope slides covered with 24×50 mm coverslips and placed in a humid box. To minimize unspecific background, do not let the slides dry during the entire procedure.

1. Remove the rubber cement frame with forceps and let the coverslip fall off.
2. Wash slides in 2× SSC for 2 min.
3. Wash slides three times in 50% or 20% formamide, 5 min each wash (see Note 6).
4. Wash slides in 2× SSC for 2 min. If fluorochrome-labeled probes are used, proceed with step 14.
5. Wash slides in 4T buffer for 5 min.
6. Incubate slides in 100 μL of blocking solution for 30 min.
7. Wash slides in 4T buffer 2×5 min.
8. Incubate slides in 100 μL of avidin–Texas Red in TNB buffer (1:1,000) for 30 min.
9. Wash slides in TNT buffer 2×5 min.
10. Incubate slides in 100 μL of mixed goat anti-avidin–biotin antibody (1:200) and mouse anti-digoxigenin antibody (1:250) in TNB buffer for 30 min (see Note 17).
11. Repeat step 9.
12. Incubate slides in 100 μL of mixed avidin–Texas Red (1:1,000) and goat anti-mouse Alexa Fluor 488-coupled antibody (1:200) in TNB buffer for 30 min.
13. Repeat step 9.
14. Dehydrate the slides in an ethanol series (70%, 80%, and 96% ethanol, 3 min each) and leave them to air-dry.
15. Apply 15 μL of Vectashield with DAPI to each slide and cover it with a 24×32 mm coverslip.
16. Observe and photograph slides under a fluorescence microscope equipped with appropriate optical filters, CCD camera, and image acquisition software.

4   Notes

1. Carnoy's II is believed to be more suitable than the Carnoy's I fixative for fixation of floral material. However, we did not observe a significant difference between the two fixatives and use the 3:1 fixative routinely.

2. The enzyme mixture can be reused several times (store at –20°C). Digestion time might need to be increased after each use.

3. Glacial acetic acid can be used at concentrations from 45% to 60%.

4. A Coplin jar can hold up to nine slides, Hellendahl jar up to 15 slides.

5. Labeled nucleotides are available commercially or can be synthesized by coupling allylamine-dUTP to succinimidyl-ester derivatives of haptenes or fluorochromes (10).

6. 50% formamide should be used for stringent post-hybridization washing after same-species (homologous) in situ hybridization; 20% formamide should be used in case of cross-species (homeologous) in situ hybridization.

7. Fixed material can be used after several months or years of storage. However, best quality preparations are usually obtained from freshly fixed material.

8. The amount of enzyme mixture should be proportional to the amount of digested material (tissue should be submerged). The incubation time of 3 h may not be appropriate to all species/types of floral material. After the incubation, we make a test preparation following protocol (3.2, steps 5–8). If the tissue is hard to disintegrate and there is a large quantity of cytoplasm and tissue fragments, we extend enzyme incubation for another 30 min or longer.

9. Digested floral tissue can be stored in 1× citrate buffer at 4°C overnight or longer. Overnight storage further softens the digested material.

10. The amount of acetic acid can be increased to 40–60 μL and the duration of suspension spreading can be prolonged. The duration of spreading can be shortened in case of very small flower buds or anthers. The needle should not touch the slide surface.

11. Post-fixed slides can be stored at 4°C for several weeks or months. However, freshly prepared or few-days old slides guarantee better results.

12. Omitting the RNase treatment may result in a longer pepsin treatment used in Subheading 3.4, step 4 and/or increased background.

13. As *Brassicaceae* possess very small genomes and low amount of repeats localized in pericentromeric heterochromatin, chromosome-specific BAC clones from euchromatic regions can be used without the need to block repetitive sequences from cross-hybridization. For chromosome painting in other plant taxa, BAC clones have to be screened for the presence of dispersed repeats or genomic/Cot DNA needs to be applied for blocking.

14. DNA of large-insert clones (usually BAC clones) can be isolated by use of a standard alkaline lysis protocol (11) or using a DNA isolation kit (e.g., the Qiagen Plasmid Midi Kit; Qiagen, Hilden, Germany). A larger amount (4 µL) of custom-made x-dUTPs must be used as custom dUTPs label DNA less efficiently than commercial nucleotides (10). DNA of several probes/BAC clones can be pulled and labeled together in a single nick translation reaction. However, in our laboratory we label each probe individually.

15. Time needed to resuspend the probe is dependent on the initial amount of precipitated DNA. DNA of several BAC clones is well dissolved after several minutes or hours. Complex probes comprising several dozen or hundred clones should be incubated overnight.

16. Overnight incubation is usually sufficient for in situ hybridization of homologous (same species) sequences. Longer hybridization times (48–72 h) may be required to ensure hybridization of homeologous (cross-species) sequences.

17. If the first detection step for hapten-labeled probes (see Subheading 3.6, step 8) yields a sufficiently strong signal, further signal amplification (e.g., by goat anti-avidin–biotin antibody in this protocol) can be omitted.

Acknowledgments

We thank Dr A. Pecinka for providing seeds of the Sha × Col hybrid. This work was supported by grants IAA601630902 and P501/10/1014 from the Grant Agency of the Czech Academy of Science and the Czech Science Foundation (GA CR), respectively. German Science Foundation (DFG) and Alexander v. Humboldt Foundation are acknowledged for supporting our research (2003–2009).

References

1. Schubert I, Fransz PF, Fuchs J, de Jong JH (2001) Chromosome painting in plants. Methods Cell Sci 23:57–69

2. Lysak MA, Fransz PF, Ali HBM, Schubert I (2001) Chromosome painting in *Arabidopsis thaliana*. Plant J 28:689–697

3. Lysak M, Fransz P, Schubert I (2006) Cytogenetic analyses of Arabidopsis. In: Salinas J, Sanchez-Serrano JJ (eds) Methods in molecular biology, vol 323, Arabidopsis protocols. Humana, Totowa, NJ, pp 173–186

4. Szinay D, Chang S-B, Khrustaleva L, Peters S, Schijlen E, Bai Y et al (2008) High-resolution chromosome mapping of BACs using multicolour FISH and pooled-BAC FISH as a backbone for sequencing tomato chromosome 6. Plant J 56:627–637

5. Febrer M, Goicoechea JL, Wright J, McKenzie N, Song X, Lin J et al (2010) An integrated physical, genetic and cytogenetic map of *Brachypodium distachyon*, a model system for grass research. PLoS One 5:e13461

6. Lysak M, Berr A, Pecinka A, Schmidt R, McBreen K, Schubert I (2006) Mechanisms of chromosome number reduction in *Arabidopsis thaliana* and related Brassicaceae species. Proc Natl Acad Sci USA 103:5224–5229

7. Ziolkowski PA, Kaczmarek M, Babula D, Sadowski J (2006) Genome evolution in Arabidopsis/*Brassica*: conservation and divergence of ancient rearranged segments and their breakpoints. Plant J 47:63–74

8. Mandáková T, Joly S, Krzywinski M, Mummenhoff K, Lysak MA (2010) Fast diploidization in close mesopolyploid relatives of *Arabidopsis*. Plant Cell 22:2277–2290

9. Lysak MA, Mandáková T, Lacombe E (2010) Reciprocal and multi-species chromosome BAC painting in crucifers (*Brassicaceae*). Cytogenet Genome Res 129:184–189

10. Henegariu O, Bray-Ward P, Ward DC (2000) Custom fluorescent nucleotide synthesis as an alternative method for nucleic acid labeling. Nat Biotechnol 18:345–348

11. Sambrook J, Russell DW (2001) Molecular cloning: a laboratory manual. Cold Spring Harbor Laboratory Press, Cold Spring Harbor, NY

12. Kocsis E, Trus BL, Steer CJ, Bisher ME, Steven AC (1991) Image averaging of flexible fibrous macromolecules: the clathrin triskelion has an elastic proximal segment. J Struct Biol 107:6–14

Using Sequential Fluorescence and Genomic In Situ Hybridization (FISH and GISH) to Distinguish the A and C Genomes in *Brassica napus*

Elaine C. Howell and Susan Armstrong

Abstract

We have developed a sequential procedure with fluorescence in situ hybridization (FISH) and genomic in situ hybridization (GISH) that enables us to distinguish between the A and C genomes in *Brassica napus* and to identify certain individual chromosomes or chromosome groups within a genome. Our modified GISH technique uses a repetitive sequence in addition to the whole genome in the blocking DNA, and it is effective on meiotic and mitotic cells present in the anther material that we use.

Keywords *Brassica*, A genome, C genome, Genomic in situ hybridization, GISH, Fluorescence in situ hybridization, FISH, Meiosis, BAC

1 Introduction

The agriculturally important crop species *Brassica napus* (oilseed rape) is an allopolyploid with the genome constitution AACC. The gene-pools of the two extant diploids, *B. rapa* (AA) and *B. oleracea* (CC), which represent the progenitor genomes, contain characteristics which are potentially useful to oilseed rape breeders. However, when a synthetic *B. napus* is produced, by hybridization between the diploids and subsequent chromosome doubling, meiotic pairing, and recombination occur between homoeologous A and C chromosomes. As a result, many of the gametes have unbalanced chromosome complements. This also happens in cultivated *B. napus*, but at a much lower frequency. We wished to investigate the extent of homoeologous pairing in male meiosis in doubled haploid (DH) lines produced from a cross between a DH line from a cultivar and a synthetic line, with a view to genetically map the locus or loci controlling this character. Therefore, we developed

Wojciech P. Pawlowski et al. (eds.), *Plant Meiosis: Methods and Protocols*, Methods in Molecular Biology, vol. 990, DOI 10.1007/978-1-62703-333-6_3, © Springer Science+Business Media New York 2013

this method using fluorescence in situ hybridization (FISH) followed by genomic in situ hybridization (GISH) (1).

With GISH, the total genomic DNA from one parental species is labeled to use as a probe. Usually an excess of DNA from the other parent is included as blocking DNA, which hybridizes to (and blocks) its own genome and sequences in the other genome that are common to both species. The labeled DNA can then hybridize only to sequences that are specific to its own genome. Since the *Brassica* A and C genomes are very closely related, it has previously been thought that there would be insufficient genome-specific sequences for GISH. However, it is apparent that the C genome has specific sequences spread widely across the genome, as we detect hybridization signals over the majority of the C chromosomes, apart from the telomeres and small subtelomeric regions, when using labeled C genome DNA. We dampen strong signals, which occur at locations corresponding to the C genome 45S rDNA sites, by adding an excess of unlabeled PCR product of the C genome 45S rDNA intergenic spacer, to produce a more even distribution of signal intensity.

Although GISH alone reveals homoeologous interactions, we find that some of the DH lines are aneuploid and our ability to interpret the pairing behavior is greatly enhanced by using both FISH and GISH. We use two FISH probes, which enable us to differentiate several groups of chromosomes and some individual chromosomes by the presence, absence, position, or size of the signals within our material. Recently, BACs (2, 3) and a C-genome-specific repeat (4) have been used as FISH probes to label the C genome. However, in situations where chromosome coverage is important, such as identifying homoeologous exchanges, GISH would be the method of choice, as it is unlikely that a BAC will contain all of the C-genome-specific sequences.

In this chapter, we describe the preparation of probes and blocking DNA, the collection of *Brassica* anthers, the preparation of slides and the FISH and GISH procedures.

2 Materials

2.1 Plant Material

Sow seeds in soil-based compost in 6 cm diameter pots and transfer plants to larger pots when necessary. Plants of *B. oleracea* var. *alboglabra* DH line A12DHd, *B. rapa* cultivar Golden Ball, *B. napus* DH12075, and other *B. napus* DH lines produce buds after about 7 weeks when grown in growth chambers maintained at 18°C with a 16 h light cycle. Sow seeds of *B. oleracea* var. *alboglabra* and *B. rapa* for DNA extraction from young leaves of young plants.

<table>
<tr><td valign="top">

2.2 Probes and Blocking DNA

</td><td valign="top">

1. FISH probes: culture *E. coli* strains and extract DNA of BAC and plasmid clones using standard methods. We routinely use two repetitive probes: DNA from *Brassica* BAC BoB061G14, which hybridizes to the pericentromeric heterochromatin of 14 of the 19 chromosome pairs (see http://www.brassica.info for available *Brassica* BACs) and DNA from clone pTa71 (EMBL X07841) (5), which hybridizes to the 45S rDNA sites (see Note 1). Each of the two probes hybridizes to both *Brassica* genomes (Fig. 1).

2. Nick translation kits for probe labeling (e.g., the DIG nick translation mix and biotin nick translation mix from Roche Applied Science, Penzberg, Germany).

3. DNA extraction kit suitable for extracting DNA from leaves. RNase may be required, depending on the kit used.

4. Agarose gel electrophoresis equipment and Lambda-phage DNA standards for DNA quantification.

5. Autoclave to fragment *B. rapa* DNA.

</td></tr>
</table>

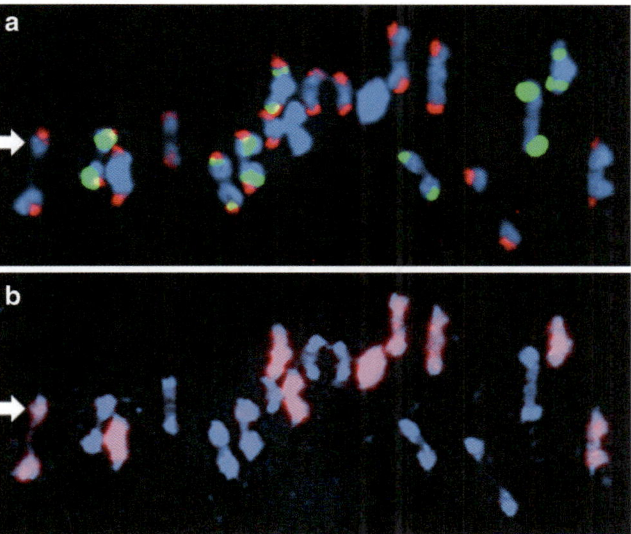

Fig. 1 Sequential FISH (**a**) and GISH (**b**) on a metaphase I spread from a DH line produced from a cross between a normal and a synthetic *B. napus*. (**a**) FISH with the 45S rDNA probe (*green*), which hybridizes to seven bivalents and BoB061G14 (*red*) which hybridizes to pericentromeric heterochromatin of 14 bivalents. (**b**) GISH showing the nine bivalents of the C genome labeled (*red*) and the ten bivalents of the A genome unlabeled. All chromosomes are counterstained with DAPI. Certain chromosomes and chromosome groups within each genome can be distinguished by comparing the images. Pairing is normal with 19 bivalents, but one bivalent (*arrowed*) is heteromorphic for one chromosome arm

6. Syringe and needle (e.g., Microlance 21G1.5 0.8×40nr2, BD; Franklin Lakes, NJ, USA) for mechanically shearing *B. oleracea* DNA.

7. PCR primers: Fwd: CAGCCCTTTGTCGCTAAG (located within the 25S gene) and Rev: GGCAGGATCAACCAGGTA (located within the 18S gene) based on the EMBL X60324 clone sequence, to amplify the intergenic spacer of *B. oleracea* var. *alboglabra* 45S rDNA.

8. PCR kit and PCR purification kit.

2.3 Fixation of Anthers and Preparation of Slides

1. Two pairs of fine watchmaker's forceps.

2. Lacto-propionic orcein stain: add 5 g of orcein (synthetic) to a mixture of 50 ml of propionic acid and 50 ml of lactic acid and stir overnight. Filter through Whatman paper (No 1). Dilute 1:1 with water for a working solution.

3. Fixative: mix 3 parts of absolute ethanol and 1 part of glacial acetic acid. Prepare fixative as required and keep on ice. Discard the unused remainder at the end of the day.

4. 0.01 M citrate buffer: to prepare 10 ml of buffer (pH 4.5) mix 445 μl of 0.1 M sodium citrate, 555 μl of 0.1 M citric acid, and 9 ml of sterile deionized water. Keep the 0.1 M sodium citrate and 0.1 M citric acid solutions at 4°C but make the buffer when required. The buffer may be stored at 4°C for a few days.

5. Stock digestion medium: dissolve 1% cellulose (Sigma Aldrich, St. Louis, MO, USA) and 1% pectolyase (Sigma Aldrich, St. Louis, MO, USA) in 0.01 M citrate buffer, pH 4.5. Dispense in aliquots of 300 μl and store at –20°C.

6. Digestion medium: mix 300 μl of stock digestion medium with 700 μl of 0.01 M citrate buffer, pH 4.5 to obtain 1 ml of 0.3% cellulose, 0.3% pectolyase in 0.01 M citrate buffer.

7. Moist box, e.g., plastic sandwich box with damp paper towels on the bottom to create a humid atmosphere. Rest slides on two glass rods or pipettes in the box.

8. 60% acetic acid: dilute glacial acetic acid with sterile deionized water.

9. Hotplate set at 45°C.

2.4 In Situ Hybridization

2.4.1 Slide Pretreatment

1. 2× SSC buffer: 0.3 M NaCl, 0.03 M tri-sodium citrate, pH 7.0. Prepare a stock solution of 20× SSC (dissolve 175 g NaCl, 88.2 g tri-sodium citrate in sterile deionized water, make up to 1 L with deionized water) and store at room temperature. Dilute to 2× SSC as required; the staining jars for eight slides that we use hold ~50 ml and this solution is required twice.

2. Pepsin digestion medium: 0.01 g pepsin in 100 ml of 0.01 M HCl. Prepare shortly before use and preheat for a few minutes at 37°C.

3. 4% Paraformaldehyde fixative: pre-warm 400 ml of sterile deionized water and ten drops of 1 M NaOH to ~60°C in a microwave oven. In a fume hood, weigh 16 g paraformaldehyde (electron microscopy grade), add it to the water and stir on a magnetic stirrer until it dissolves. Filter through Whatman paper. Allow to cool to room temperature before adjusting to pH 8.0 with 1 M HCl. Store at 4°C.

4. Alcohol series: 70% and 85% absolute ethanol in sterile deionized water and 100% absolute ethanol.

2.4.2 Denaturation, Hybridization, and Post-hybridization Treatments

1. Hybridization mix: add 5 ml of formamide (molecular biology grade) and 1 ml of 20× SSC to 1 g dextran sulfate (high MW 500,000) and make up to 7 ml with sterile deionized water. Dissolve at 65°C and adjust to pH 7.0 after cooling down. Aliquot into 1 ml microfuge tubes and store at –20°C. For one slide, 20 µl of probe mixture is used, which contains 14 µl of hybridization mixture and, therefore, the final concentrations in the probe mixture are 50% formamide, 2× SSC, and 10% dextran sulfate pH 7.0.

2. Labeled FISH probes (see Subheading 2.2, items 1 and 2, and Subheading 3.1, step 1) or labeled probe and unlabeled blocking DNA for GISH (see Subheadings 2.2 and 3.1).

3. Vulcanizing rubber solution (e.g., for repairing bicycle tires).

4. Hotplate set at 75°C.

5. Moist box (see Subheading 2.3, item 7).

6. Water bath set at 45°C.

7. Post-hybridization washes: three staining jars with 50% deionized formamide, 2× SSC, pH 7.0. We use 200 ml of formamide (high-purity grade) with 10 g mixed bed resin (Amberlite, Dow Chemical, Midland, MI, USA) stirred for 1 h in a fume hood and filtered. Mix 150 ml of this deionized formamide with 30 ml of 20× SSC and 120 ml of sterile deionized water. Warm to 45°C in a water bath and adjust to pH 7.0 with 1 M HCl, being careful not to go below this pH. Allow to cool and pour into three staining jars. Also, make up one staining jar of 2× SSC and one staining jar of 4× SSC, 0.05% Tween 20. Store the five jars at 4°C between uses and replace solutions monthly. Make up a solution of 4× SSC, 0.05% Tween 20 (to make 1 L, mix 200 ml of 20× SSC, 800 ml of sterile deionized water, and 0.5 ml of Tween 20) for all subsequent washes on the day.

8. Detection of labeled probes: a digoxigenin-specific antibody conjugated to a fluorophore, e.g., anti-digoxigenin-fluorescein,

and a product for biotin detection, such as Cy3 Streptavidin. Prepare a digoxigenin blocking solution (0.5 g Roche DIG blocking reagent in 100 ml 4× SSC, 0.05% Tween 20) and a biotin blocking solution (5 g dried skimmed milk in 100 ml 4× SSC, 0.05% Tween 20). Centrifuge the solutions, aliquot the supernatants into 1 ml microfuge tubes and freeze.

9. Parafilm.

10. DAPI: prepare a stock solution of 4′,6-diamidino-2-phenylin-dole (DAPI) in sterile deionized water at 1 mg/ml, dispense in aliquots and store at –20°C. For use, add 10 µl of this stock to 1 ml of an anti-fade mounting medium, Vectashield (Vector Laboratories, Berlingame, CA) and store at 4°C.

11. Fluorescence microscope with appropriate filters, a motorized stage, and an image capture and analysis system (see Note 2).

2.4.3 Re-probing Slides

1. All materials in Subheading 2.4.2, including the GISH probe and blocking DNA but not the FISH probes, and only biotin detection reagents.

2. 2× SSC in a staining jar.

3. Alcohol series: 70% and 85% absolute ethanol in sterile deionized water and 100% absolute ethanol.

3   Methods

3.1 Preparing Probes and Blocking DNA

1. FISH probes: label DNA of the BoB061G14 BAC clone with biotin-16-dUTP and DNA of 45S rDNA repeat with digoxigenin-11-dUTP by nick translation following manufacturer's instructions.

2. Extract DNA from young leaves of *B. oleracea* var. *alboglabra* and also from *B. rapa* and treat it with RNase.

3. Quantify the DNA by agarose gel electrophoresis with Lambda-phage DNA standards. The concentration of the *B. rapa* DNA, which will be used for blocking, should be at least 275 ng/µl.

4. Transfer the *B. rapa* DNA to screw-capped tubes (we put 150 µl of the DNA in each tube) and autoclave for 5 min at 15 psi. Run samples on a gel to determine the fragment size. Repeat the autoclave and gel steps to obtain fragments of 100–500 bp. In our hands, autoclaving twice is sufficient. If evaporation occurs, remeasure the volumes and recalculate the DNA concentrations.

5. Keep a small volume of *B. oleracea* DNA for a PCR analysis. Perform PCR with primers listed in Subheading 2.2, item 7, using an annealing temperature of 61°C to amplify the intergenic spacer of the 45S rDNA repeat. Clean the product

with a PCR purification kit. This solution will be used for blocking.

6. Shear the remainder of *B. oleracea* DNA mechanically by repeatedly (~20 times) drawing in and out through a syringe needle. Label this DNA by nick translation with biotin-16-dUTP when required.

3.2 Fixation of Anthers and Preparation of Slides

1. Fixation of Anthers: place young terminal inflorescences on damp filter paper in a petri dish. Using a stereomicroscope and two pairs of fine watchmaker's forceps, dissect out individual buds and place them in size order on the filter paper, keeping each inflorescence separate. Alternatively, take buds directly from the plant onto damp filter paper, leaving smaller buds to develop. The pollen mother cells in anthers of buds approximately 0.5–2 mm in length will be the ones undergoing meiosis. Tapetal cells will be in mitosis in buds at the lower end of this size range. Remove one anther from a bud, place it on a microscope slide with lacto-propionic orcein stain, tap with a brass rod, add a coverslip, and examine with a phase-contrast microscope to determine the meiotic stage. Place the remaining anthers in fresh, ice-cold fixative in a microfuge tube. Store anthers from several buds at the same stage in the same tube. Leave at room temperature and replace the fixative after 2–3 h. Use after 24 h or store at 4°C for up to 6 months.

2. Digestion: place fixed anthers and fixative into a watch glass (preferably black). Using Pasteur pipettes, replace the fixative with 0.01 M citrate buffer. Wash with the same buffer 3×2 min. Replace the citrate buffer with digestion medium (we use ~300 µl for one watch glass with ~20 anthers). Incubate in a moist box at 37°C for 60–120 min depending on anther size. Replace the digestion medium with ice-cold citrate buffer to stop digestion.

3. Slide preparation: Place a single anther on a slide (with frosted end), with a minimum of liquid. Macerate the anther with a mounted needle quickly without letting the material dry. Add 10 µl of 60% acetic acid. With a diamond pen, mark a region around the sample, no greater than the size of a 22×22 mm coverslip. Place the slide on a hot-block at 45°C for about 10 s and during this time add another 10 µl of 60% acetic acid and mix gently once with the mounted needle. Remove the slide and add 100 µl of ice-cold fixative as a ring around the material. Tip the slide to drain it and dry gently with a hair drier directed onto the back of the slide.

4. Examine the slides with a phase-contrast microscope to select those with good quality spreads of the required meiotic or mitotic stage. Slides are usually kept overnight at room temperature before in situ hybridization.

3.3 In Situ Hybridization

3.3.1 Slide Pretreatment

1. Place five to eight slides in an appropriate staining jar.
2. Wash the preparations in 2× SSC for 10 min at room temperature.
3. Digest with pepsin digestion medium at 37°C for 1–2 min.
4. Wash in 2× SSC briefly.
5. Fix the material in 4% paraformaldehyde in a fume hood for 10 min.
6. Wash in sterile deionized water briefly.
7. Dehydrate the preparations through an alcohol series (70%, 85%, and 100% ethanol for 2 min each).
8. Remove the slides from the staining jar and support them vertically on tissues to drain dry.
9. Check the material is still present using a phase-contrast microscope.

3.3.2 FISH (See Note 3)

1. Prepare a probe mixture: use 14 µl hybridization mix, 0.5–2 µl of each FISH probe and sterile deionized water to 20 µl, per slide.
2. Denaturation: apply 20 µl of probe mixture to each slide. Add a coverslip (22×22 mm) and seal the edges with vulcanizing rubber solution. Place on a hotplate at 75°C for 3 min 30 s.
3. Hybridization: place the slides in a moist box at 37°C overnight.
4. Heat the first 5 post-hybridization washing solutions in a water bath at 45°C. Remove the rubber solution and slide off the coverslips using fine forceps. Place the slides in the first wash of 50% deionized formamide, 2× SSC, pH 7.0 for 5 min. Transfer to two further washes of the same solution, then to 2× SSC, and then to 4× SSC, 0.05% Tween 20 at 5 min intervals.
5. Transfer the slides to a staining jar with 4× SSC, 0.05% Tween 20 at room temperature for 5 min.
6. During these washes, prepare the detection solutions: use anti-digoxigenin-fluorescein in digoxigenin blocking solution (5 ng/µl) and Cy3-streptavidin in biotin blocking solution (5 ng/µl) or as manufacturer's instructions. Prepare 100 µl of each solution per slide and keep on ice in the dark.
7. Drain a slide (but do not dry), add 100 µl of the anti-digoxigenin detection solution, and place a piece of parafilm, cut to the size of the slide, over the solution. Repeat with the remaining slides and place in a moist box at 37°C in the dark for 30 min.
8. Remove the parafilm and wash slides 3×5 min in 4× SSC, 0.05% Tween 20 at room temperature in the dark.
9. Repeat steps 7 and 8 with 100 µl of the Cy3-streptavidin detection solution.

10. Drain each slide (but do not dry), add 7 μl of DAPI and a coverslip. Keep in the dark at 4°C.

11. View slides with a fluorescence microscope with the appropriate filters, an automated stage, and an image capture and analysis system (Fig. 1). Note the exact positions of the imaged cells so that you can reimage the same cells in Subheading 3.3.3, step 10.

3.3.3 Re-probing Slides, GISH (See Notes 2 and 3)

1. Place slides in a staining jar with 2× SSC for approximately 5 min until the coverslips slide off easily.

2. Dehydrate slides through an alcohol series (70%, 85%, and 100% ethanol for 1–2 min each) and leave to dry.

3. Prepare 20 μl of GISH probe mixture per slide: 14 μl of hybridization mix, 1 μl (50 ng) of biotin-labeled *B. oleracea* DNA, 4 μl (1,100 ng) of fragmented *B. rapa* DNA, and 1 μl of PCR product of the *B. oleracea* 45S rDNA intergenic spacer (see Note 4).

4. Denaturation: apply 20 μl of the probe mixture to each slide. Add a coverslip (22 × 22 mm) and seal the edges with vulcanizing rubber solution. Place on a hotplate at 75°C for only 3 min to denature probe and chromosomes and to release previously bound probe (see Note 5).

5. Follow Subheading 3.3.2, steps 3–5.

6. Prepare the detection solution: Cy3-streptavidin in biotin blocking solution (5 ng/μl) or as *per* manufacturer's instructions. Prepare 100 μl per slide and keep on ice in the dark.

7. Drain a slide, add 100 μl of detection solution, and cover with a piece of parafilm. Repeat with the remaining slides. Place the slides in a moist box at 37°C in the dark for 30 min.

8. Remove the parafilm and wash slides 3 × 5 min in 4× SSC, 0.05% Tween 20.

9. Drain each slide and add 7 μl of DAPI and a coverslip. Keep in the dark at 4°C.

10. View slides with the fluorescence microscope and recapture the same spreads previously imaged in Subheading 3.3.2, step 11 (Fig. 1).

---

4 Notes

1. Because both probes hybridize to repetitive sequences, we do not include C_0t-1 DNA (6) as this would block these sites. However, if we use genetically mapped *B. oleracea* BACs to identify specific chromosomes, we include C_0t-1 DNA to suppress non-specific signals (1).

2. Performing FISH and GISH sequentially requires a fluorescence microscope equipped with a motorized stage and an image capture and analysis system that automatically returns the stage to the coordinates of captured spreads.

3. If FISH and GISH are used sequentially, performing FISH first is preferable as the spreads deteriorate a little on re-probing.

4. Since the DNA concentrations are not quantified very accurately, preliminary tests may be required with different volumes of probe and block in the probe mixture, or different concentrations of block. If there is insufficient blocking DNA, more labeled DNA hybridizes to pericentromeric regions of A genome chromosomes and to the intergenic spacer regions of 45S rDNA sites on C genome chromosomes.

5. If performing GISH without FISH, leave out Subheadings 3.3.2 and 3.3.3, steps 1 and 2 and increase the denaturation time at 75°C to 3 min 30 s. Remember that Subheading 3.3.2, steps 3–5 are still required for hybridization, and post-hybridization washing of the GISH slides.

Acknowledgments

E.C. Howell was supported by BBSRC. We thank S. Price for technical assistance and I.A. Parkin (AAFC, Saskatoon, Canada) for *B. napus* seed.

References

1. Howell EC, Kearsey MJ, Jones GJ, King GJ, Armstrong SJ (2008) A and C genome distinction and chromosome identification in *Brassica napus* by sequential fluorescence in situ hybridization and genomic in situ hybridization. Genetics 180:1849–1857

2. Leflon M, Eber F, Letanneur JC, Chelysheva L, Coriton O, Huteau V et al (2006) Pairing and recombination at meiosis of *Brassica rapa* (AA) x *Brassica napus* (AACC) hybrids. Theor Appl Genet 113:1467–1480

3. Xiong ZY, Pires JC (2011) Karyotype and identification of all homoeologous chromosomes of allopolyploid *Brassica napus* and its diploid progenitors. Genetics 187:37–49

4. Alix K, Joets J, Ryder CD, Moore J, Barker GC, Bailey JP et al (2008) The CACTA transposon Bot1 played a major role in *Brassica* genome divergence and gene proliferation. Plant J 56:1030–1044

5. Gerlach WL, Bedrock JR (1979) Cloning and characterization of ribosomal RNA genes from wheat and barley. Nucleic Acids Res 7:1869–1885

6. Zwick MS, Hanson RE, Islam-Faridi MN, Stelly DM, Wing RA, Price HJ et al (1997) A rapid procedure for the isolation of C_0t-1 DNA from plants. Genome 40:138–142

Chapter 4

Labeling Meiotic Chromosomes in Maize with Fluorescence In Situ Hybridization

Zhi Gao, Fangpu Han, Tatiana V. Danilova, Jonathan C. Lamb, Patrice S. Albert, and James A. Birchler

Abstract

Fluorescence in situ hybridization (FISH) can be used to visualize chromosomal features using repetitive or single gene probes above a minimum target size. When applied to meiosis, each chromosome of the karyotypic complement can be identified, which can facilitate an understanding of the interrelationship of different chromosomes during this process. On the other hand, the pachytene stage of early meiosis is characterized by slightly but not strongly condensed chromosomes that permit more detailed analyses of adjacent features than can be achieved with somatic metaphase chromosomes.

Keywords FISH, Chromosomes, Pachynema, Meiosis

1 Introduction

The application of fluorescence in situ hybridization (FISH) to meiotic cells has allowed a precise analysis of chromosomal aberrations as well as the behavior of chromosomes. A cocktail of clustered repetitive sequences was used in maize to identify the ten pairs of homologues in somatic root tip chromosome preparations as well as in meiotic stages, in which the different chromosomes could not otherwise be distinguished (1) (Fig. 1). Subsequently, a technique was developed that could visualize single genes with a target size at a minimum of about 2–2.5 kb (2, 3). Also, identification of unique sequences within bacterial artificial chromosomes (BACs) allowed predetermined sites on chromosomes to be visualized by FISH (2, 4). All of these procedures can be applied to meiosis where the potential applications of FISH are numerous. The mechanics of chromosome movement can be detailed for each chromosome (1, 5). The generalized distribution of transposons can be visualized (6). Various types of chromosomal aberrations of

Wojciech P. Pawlowski et al. (eds.), *Plant Meiosis: Methods and Protocols*, Methods in Molecular Biology, vol. 990, DOI 10.1007/978-1-62703-333-6_4, © Springer Science+Business Media New York 2013

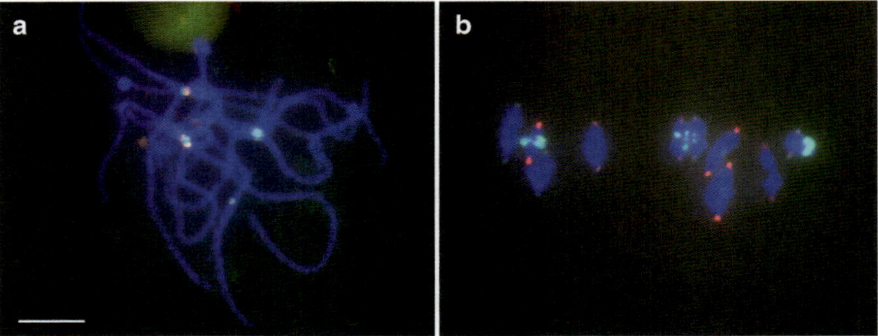

Fig. 1 FISH applied to different stages of meiosis in maize. (**a**) Pachytene stage of meiosis from inbred line B73. Chromosomes are stained with DAPI and are *blue*. The centromeric satellite, CentC, is labeled in *red* and the centromeric retrotransposon (CRM) in *green*. The *diffuse green sphere* at the top is the nucleolus. (**b**) Metaphase I of meiosis from inbred line B73. Chromosomes are labeled in *blue*. CentC is in *red*. CRM and knob heterochromatin are labeled in *green*. Bar = 10 μm

complex structure can be deciphered using specially designed probes (7, 8). FISH mapping also can be used for studying chromosomal regions with suppressed recombination (9). These citations provide only a sampling of the possible applications of FISH for meiotic study.

A previous publication in this series has detailed the methods of probe preparation, probe labeling, hybridization, and various other considerations about optimizing FISH (10). Because these methods are identical to those used for meiosis, they will not be reiterated here. In this chapter, we focus on aspects of FISH that are specific to meiosis.

2 Materials

2.1 Collection and Fixation of Plant Material

1. Plants at proper stage for harvesting immature tassels (see Note 1).
2. Razor blade or scalpel.
3. 50-ml conical tubes (one per tassel).
4. Bucket of ice.
5. Tassel fixative: 95% ethanol:glacial acetic acid (3:1 v/v). Make just before use and chill at 4°C or on ice.
6. 70% ethanol solution: dilute from 95% ethanol.

2.2 Preparation of Meiotic Spreads for Staging of Anthers

1. Fixed immature tassels.
2. Scalpel and two pairs of tweezers (medium point for spikelets; fine, for anthers).
3. One or two large petri dishes (for working with tassels).

4. Squeeze bottle filled with 70% ethanol.

5. Bucket of ice.

6. 0.5-ml microcentrifuge tubes for storage of staged anthers.

7. Glass slides and glass 22 × 22 mm coverslips.

8. Aceto carmine stain: purchase or make your own (see Chapter 8, Subheading 2.2, item 2 for a detailed protocol).

9. Rusty dissecting probe (presence of iron potentiates aceto-carmine staining).

10. Alcohol lamp and lighter.

11. Compound light microscope with 10× and 40× objective lenses.

2.3 Preparation of Meiotic Spreads for Hybridization

The reagent and supply list below is slightly modified from a previous publication describing mitotic cell FISH (10).

1. 5× citric buffer: 50 mM sodium citrate, 50 mM EDTA, adjusted to pH 5.5 by adding citric acid. Autoclave and store at room temperature. Dilute to 1× with sterile water. Use chilled on ice.

2. Digestive enzyme solution (see Note 2): 0.1 g (1% w/w) of pectolyase Y-23, 0.2 g (2% w/w) of cellulase Onozuka R-10, and 9.7 g 1× citric buffer. Makes ~10 ml. Mix thoroughly in tube or weigh boat on ice. Quickly dispense as 20 μl aliquots into thin-walled 0.5-ml PCR tubes (make sure that caps are tightly closed), quick-freeze on dry ice, and store at –20°C. The stock is good for about 1 year.

3. 37°C water or dry bath for tissue digestion.

4. 10× TE buffer stock: 100 mM Tris base, 10 mM EDTA adjusted to pH 7.6 with HCl. Pass through a 0.45-μm filter and autoclave. Prepare working solution of 1× TE by diluting with sterile water. Also needed is a 1× working stock at pH 7.8 for preparation of salmon sperm DNA stock (see Subheading 2.4, item 7). Store at room temperature.

5. 100% ethanol.

6. Cell suspension solution: 100% acetic acid or, alternatively, a solution containing 1 volume of methanol and 3–9 volumes of acetic acid, freshly prepared.

7. Humid chamber for spreading of cells (see Note 3).

8. UV Cross-linker that can deliver a total energy of 120–125 mJ/cm^2 at "Optimum" cross-link setting.

9. Miscellaneous supplies: microcentrifuge tubes, tweezers, scalpel, dissecting needle with the point ground off, microscope slides.

2.4 Denaturation, Hybridization, and Slide Processing

The reagent and supply list below is slightly modified from a previous publication describing mitotic cell FISH (10).

1. Probe cocktail(s) and cross-linked slides.

2. Boiling water bath: use a hot plate and a metal baking pan that holds boiling water (see Note 4).

3. Denaturation pan assembly: a smaller baking pan (aluminum, if possible) that can float freely in the water bath and a plastic pipet tip box lid (or similar) to cover the slides/probes in the pan. The bottom of the baking pan is covered with two to three layers of paper tissue that are sprayed with purified water just before use. The water bath and floating pan are covered with aluminum foil to retain heat and block light during the denaturation step.

4. Ice bucket filled to the top with crushed ice and a small beaker filled with slushy ice.

5. Metal plate(s) on which to quick chill several slides simultaneously (e.g., two at $8 \times 15 \times 0.6$ mm thick). A stainless-steel slide tray, turned upside down, may be used for this purpose.

6. 20× SSC: 3 M sodium chloride, 0.3 M sodium citrate, pH 7.0 (see Note 5). Pass through a 0.45-μm filter (wet the filter with small volume of water before use), autoclave, and store at room temperature.

7. 10 mg/ml salmon sperm DNA (see Note 6).

8. 2× SSC, 1× TE containing 140 ng/μl salmon sperm DNA. Make 1 ml by diluting 20× SSC (see item 6 above), 10× TE (see Subheading 2.3, item 4), and 10 mg/ml autoclaved salmon sperm DNA (see item 7 above) with sterile water.

9. Coverslips: 22×22 mm plastic (Thermo Fisher Scientific, Pittsburgh, PA) and 24×50 mm glass, No. 1 (Corning; Thermo Fisher Scientific, Pittsburgh, PA).

10. Tweezers for handling coverslips and either a large serrated pair of tweezers or tongs for removing denaturation pan from boiling water bath.

11. Mini microcentrifuge (2,000 rcf) for quick spin of chilled, denatured probes.

12. Humidified storage container for hybridization: reusable airtight food container, bottom lined with paper tissue moistened with water.

13. Dedicated incubator/oven set to 55°C for hybridization.

14. 2× SSC: Make by diluting the 20× stock (see item 6 above) with sterile water.

15. Two Coplin jars, one at room temperature and one preheated to 55°C into which ~45 ml of preheated 2× SSC (not the buffer containing TE) will be added.

16. Antifade mounting medium: Vectashield containing 1.5 µg/ml DAPI (Vector Laboratories, Burlingame, CA) diluted 1/20 or 1/40 with DAPI-free Vectashield (see Note 7).

17. Fluorescence microscope with a 20× or 40× oil lens for scanning slides, a 100× oil lens for image acquisition (a flat field and color corrected lens is best), filter sets for fluorochromes to be used (see Note 8), cooled charge-coupled device (CCD) camera, and acquisition software.

3   Methods

3.1   Collection and Fixation of Plant Material

1. Use a razor blade to slice open the stalk a few leaves at a time (don't cut too deeply). If the tassel is ready to harvest, cut off the top of the plant below the tassel, then carefully excise the tassel, including a small portion of the stalk at the base of the tassel (see Note 9).

2. Immediately immerse tassel in approximately 40 ml of fresh, ice-cold fixative in the 50-ml conical tube. Top off with fixative to cover the whole tassel and then place the tube on ice. Harvest any remaining samples. Fix all tassels for 24 h at 4°C.

3. Rinse the tassels three times with 70% ethanol and store in 70% ethanol at –20°C.

3.2   Preparation of Meiotic Spreads for Staging of Anthers

Unless otherwise specified, the sample does not need to be kept on ice. Steps 2 and 3 may be performed on the benchtop (dark background helpful) or on the stage of a dissecting microscope.

1. Fill a petri dish with 70% ethanol, place on ice, and transfer a stored tassel to the dish.

2. Use the tweezers to remove a spikelet (note its location on the tassel) and place it in 70% ethanol in the second petri dish (or in a large drop of 70% ethanol on a microscope slide).

3. Remove the three largest anthers by one of the following methods:

    (a) While holding the spikelet steady with tweezers, use a scalpel to remove the base of the spikelet very close to the anthers and then tease apart the surrounding tissue to free the large anthers. Discard the smaller anthers.

    (b) After removing the base of the spikelet, make a cut that separates the two anther groups and then extricate the large anthers from the remaining tissue.

4. Transfer two of the anthers to a microcentrifuge tube containing 70% ethanol. Place on ice.

5. Place the third anther on a dry microscope slide and quickly add a drop of aceto carmine stain. Crush the anther with the

rusty probe to remove the meiocytes and expose them to the stain. Alternatively, cut the anther into about three equal pieces then crush with the probe.

6. Apply a coverslip. If the anther has not been reduced to small enough pieces, use the blunt end of the rusty probe to tamp lightly on the coverslip.

7. Wave the slide over, and close to, the flame of the alcohol lamp (see Note 10). Repeat two to three times. Heat thoroughly but do not boil.

8. Use a microscope to check for the desired meiotic stage (see Note 11). For a detailed description of maize microsporogenesis, see Chang and Neuffer (11).

9. If the other two anthers are to be kept, label the tube appropriately and store at −20°C. Otherwise, discard.

10. When finished, return the tassel to its storage tube and place at −20°C.

3.3 Preparation of Meiotic Spreads for Hybridization

These procedures are nearly the same as described previously for root tip metaphase spreads (10).

1. Soak anthers in ice-cold, 1× citric buffer for 5–10 min to remove residual ethanol. Alternatively, use sterile water. The anthers will probably float.

2. Place anthers in ice-cold enzyme cocktail, then transfer to 37°C for 20–60 min depending on the size of the anther and genotype (see Notes 12 and 13). Anthers should be submerged.

3. The following washes are performed on ice: the enzyme solution is rinsed away by filling the tube with cold 1× TE. The TE is removed with a pipet and the anthers are gently rinsed three times in 100% ethanol.

4. Then, 20–35 μl of acetic acid-methanol or 100% acetic acid (see Note 14) is added and the anthers broken apart with a dissecting needle. The tube is gently agitated to separate cells. About 8 μl of the cell suspension is dropped onto a glass slide in a humid chamber and allowed to dry. The dried preparations can be examined with a light microscope to confirm that the desired meiotic stage(s) are present.

5. The slides are placed in a UV cross-linker set at "Optimum" to cross-link the chromatin to the slide.

3.4 Denaturation, Hybridization, and Slide Processing

Methods for probe construction, denaturation, hybridization of probes, stringency conditions for washing, and microscopic examination are exactly the same as previously described for somatic metaphase spreads (10).

4 Notes

1. Meiosis in maize is typically examined from immature tassels, which form between 1 and 2 months after planting (environment and genotype dependent). The stages of meiosis are not synchronized through a particular tassel—a fact that allows one to examine multiple stages of the process with a single sampling. Although there is some variability, the usual circumstance for the best sampling is to slice open the whorl of leaves and remove the tassel that is completely formed but is pale in color. Plants with approximately eight fully developed leaf sheaths are good candidates for tassel harvest (12). If the anthers in the tassel exhibit a yellow color, those typically have already formed pollen and the optimal stage for collection has passed. Those tassels that are not yet fully formed usually will not yet show meiotic stages with the potential exception of the very earliest ones. If the tassel has begun to emerge from the whorl of leaves, it will be past the timing of meiosis. It is possible in maize to sample only a portion of a tassel and tape the cut leaves together again, thus allowing the remaining portion of the tassel to develop and produce pollen that can be used in genetic crosses. By this method, it is possible to examine meiosis of a specific plant and then use it in genetic crosses for subsequent analysis.

   In an appropriately sampled tassel, different stages of meiosis are present in different locations. Typically, the anthers with the most advanced stage will be present in the middle of the top spike and in the middle of the tassel branches. Earlier stages of meiosis can be sought by examining anthers from more distal or more proximal positions along the length of the branch. The last portion of the tassel with cells undergoing meiosis is at the base of the tassel.

   Each spikelet on the tassel contains two florets, one of which is more developed and has larger anthers, and each floret contains three anthers that exhibit the same meiotic stage. Thus, by determining the stage of one of the three anthers using aceto carmine staining and light microscopy, the other two can be saved for later analysis of the determined stage.

2. Because enzyme quality varies among suppliers and lot numbers, suppliers' names are not provided.

3. One version of a humid chamber consists of an open-top cardboard box (~7 cm high), subdivided into sections that will easily accommodate the width and number of microscope slides to be processed. All surfaces are lined with multiple layers of paper towels (stapled into position), which are moistened with purified water prior to use. Slides placed in the chamber are elevated (e.g., on dry plant stake or upside down tube rack).

Covering the chamber with a large paper tissue, such as Kimwipe, (either damp or dry) is optional. The chamber is reused.

4. The metal baking pan in contact with the hot plate should not be made of thin gauge aluminum, which tends to warp. A thermostatically controlled electric skillet with lid may be used in lieu of the water bath.

5. Optional addition: EDTA to a final concentration of 20 mM.

6. Dissolve DNA in 1× TE buffer pH 7.8 (see Subheading 2.3, item 4) at a concentration of 10 mg/ml. Place on a rotary shaker (overnight, if necessary). The DNA should be completely dissolved prior to autoclaving for 30 min (this treatment shears DNA). Run sample of the sheared DNA on gel to determine the size of the DNA fragments. We use fragments ~100–400 bp in length (purchased sheared DNA with fragments of 300–2,000 bp also works). Store at –20°C in 1-ml aliquots. If a white precipitate (probably undissolved DNA) is visible, centrifuge 5 min at 16,000 rcf and transfer DNA to new tube. Final concentration, determined spectrophotometrically, will not be exactly 10 mg/ml; a concentration of 7–11 mg/ml works well.

7. We have not tried the "Hard-Set" version of Vectashield.

8. If imaging red and far-red signals from the same specimen, the filter sets must be custom made to avoid bleed through of bright red signals into the far-red channel and vice versa. As less overlap is achieved, signal strengths are decreased. Consult a microscopy optics specialist for specifications.

9. If plants are growing in the field, it is convenient to cut off the tops of the plants and take them back to the laboratory for tassel excision and fixation.

10. Some lighters tend to produce soot, causing accumulation of a residue on the slide, and are not recommended for this step other than to light the alcohol lamp.

11. To better visualize the chromosomes, it sometimes helps to open the condenser aperture and also increase the light intensity at the source. One slide can be used repeatedly for viewing anther preparations; simply push the coverslip off with a wadded up lint-free paper tissue, such as Kimwipe, and rub the slide a few times to clean. Wash with water when necessary.

12. If a genotype is found to have an abundance of cytoplasm, nicking or cutting the anther prior to the digestion step increases contact with the enzyme and may produce better results.

13. The cytoplasm in meiotic preparations tends to generate greater background than is typically present with somatic chromosome

preparations. Selecting slides with minimum cytoplasm for hybridization will reduce the adherence of probe to this material and provide a better signal-to-noise ratio. For single gene detection, only slides with a minimum of cytoplasm at this stage should be used. Some procedures have described various treatments to reduce or eliminate the cytoplasmic background in order to detect small target signals. Increasing the length of incubation of anthers in the cellulase–pectolyase enzyme cocktail reduces the background, as do increasing the concentration of cellulase to 4% or resuspending the cells in 100% acetic acid.

14. If 100% acetic acid is used, the cell suspension cannot be stored; it needs to be dropped onto slides, which can then be stored for approximately a week or more at 4°C.

Acknowledgments

This work was supported by grants from the National Science Foundation DBI 0423898, DBI 0421671, and DBI 0701297.

References

1. Kato A, Lamb JC, Birchler JA (2004) Chromosome painting using repetitive DNA sequences as probes for somatic chromosome identification in maize. Proc Nat Acad Sci USA 101:13554–13559

2. Lamb JC, Danilova T, Bauer MJ, Meyer J, Holland JJ, Jensen MD et al (2007) Single gene detection and karyotyping using small target FISH on maize somatic chromosomes. Genetics 175:1047–1058

3. Kato A, Albert PS, Vega JM, Birchler JA (2006) Sensitive FISH signal detection using directly labeled probes produced by high concentration DNA polymerase nick translation in maize. Biotech Histochem 81:71–78

4. Danilova TV, Birchler JA (2008) Integrated cytogenetic map of mitotic metaphase chromosome 9 of maize: resolution, sensitivity and banding paint development. Chromosoma 117:345–356

5. Wolfgruber TK, Sharma A, Schneider KL, Albert PS, Koo D-H, Shi J et al (2009) Maize centromere structure and evolution: sequence analysis of centromeres 2 and 5 reveals a major role for retrotransposons. PLoS Genet 5:e1000723

6. Lamb JC, Meyer JM, Corcoran B, Kato A, Han F, Birchler JA (2007) Distinct chromosomal distributions of highly repetitive sequences in maize. Chromosome Res 15:33–49

7. Zhang J, Yu C, Pulletikurti V, Lamb J, Danilova T, Weber DF, Birchler J, Peterson T (2009) Alternative Ac/Ds transposition induces major chromosomal rearrangements in maize. Genes Dev 23:755–765

8. Yu C, Danilova TV, Zhang J, Birchler JA, Peterson T (2010) Constructing defined chromosome segmental duplications in maize. Cytogenet Genome Res 129:72–81

9. Szinay D, Chang S-B, Khrustaleva L, Peters S, Schijlen E, Bai Y et al (2008) High-resolution chromosome mapping of BACs using multicolour FISH and pooled-BAC FISH as a backbone for sequencing tomato chromosome. Plant J 56:627–637

10. Kato A, Lamb JC, Albert PS, Danilova T, Han F, Gao Z et al (2011) Chromosome painting for plant biotechnology. In: Birchler JA (ed) Plant chromosome engineering. Humana, New York, pp 67–96

11. Chang MT, Neuffer MG (1994) Chromosomal behavior during microsporogenesis. In: Freeling M, Walbot V (eds) The maize handbook. Springer, New York, pp 460–475

12. Li X, Topp CN, Dawe RK (2012) Maize antibody procedures: Immunolocalization and chromatin immunoprecipitation. In: Bass HW, Birchler JA (eds) Plant cytogenetics: genome structure and function. Springer, New York, pp 273–274

Chapter 5

Examining Female Meiocytes of Maize by Confocal Microscopy

Philippa Barrell and Ueli Grossniklaus

Abstract

Meiosis is a key step in sexual plant reproduction. Cytological analysis of meiosis enables hypotheses of gene function during meiosis to be confirmed or discarded. Observation of meiosis in the female reproductive organs in many higher plants is technically challenging because the cells are enclosed deep within the nucellus and ovule tissues. In this chapter, we describe a simple staining procedure using Schiff's reagent optimized for use with confocal microscopy to observe meiosis within intact ovules of *Zea mays* (maize). Using this procedure we present images taken by confocal laser scanning microscopy of wild-type meiotic cells in whole-mount maize ovules.

Keywords Confocal microscopy, Meiosis, Ovule, Schiff's reagent, Dissection

1 Introduction

The embryo sac, or female gametophyte, of flowering plants represents the multicellular haploid generation of the plant life cycle that is contained within the ovule, the female reproductive organ in the flower. The processes of megasporogenesis and megagametogenesis occur coordinately with the development of the ovule (1–4). A subepidermal cell differentiates from surrounding nucellar cells and develops into an archesporial cell, which in turn differentiates into the megaspore mother cell without additional division, and initiates meiosis.

Meiosis is a key event in sexual reproduction, where recombination and segregation take place to create new combinations of alleles. Prophase of meiosis I is divided into leptotene, when chromosomes condense and the formation of the synaptonemal complexes is initiated, zygotene, when homologous chromosomes synapse, pachytene, when synapsis is complete, diplotene, when the synaptonemal complex degrades, and diakinesis, when chromosomes are highly condensed and chiasmata are clearly visible.

Wojciech P. Pawlowski et al. (eds.), *Plant Meiosis: Methods and Protocols*, Methods in Molecular Biology, vol. 990, DOI 10.1007/978-1-62703-333-6_5, © Springer Science+Business Media New York 2013

Each of these substages is readily observable cytologically (5–7). The product of meiosis I is a pair of cells, the dyad, but in some species no cytokinesis takes place after meiosis I (5, 8, 9). In meiosis II, each of the two cells of the dyad divides once more, proceeding through metaphase and anaphase without further chromosome replication. The second division of meiosis produces a tetrad of haploid spores. Three of the spores degenerate and the remaining chalazal spore differentiates into the functional megaspore.

Numerous techniques for staining DNA within tissues have been published. However, due to the difficulties in gaining access to female gametophytic tissues, fewer publications exist describing DNA staining and observation of female meiosis. Dissection of individual female meiocytes and embryo sacs in maize has been reported (10). As an alternative approach, fixation, embedding, and sectioning of tissues has also been reported in maize (11, 12). In addition to being technically challenging, the disadvantage of dissecting and isolating individual cells is that the positioning of the cells relative to their neighbors is lost. Embedding and sectioning of tissue is time-consuming, and resulting section planes are difficult to reconstruct into single three-dimensional (3D) images.

Here, we describe Feulgen staining with Schiff's reagent, combined with confocal microscopy, suitable for observation of meiosis. The method was adapted and optimized for whole-mount ovules based on earlier protocols (10, 13). Meiosis was observed within the context of intact sporophytic tissue. 3D images can be readily reconstructed from optical section planes.

Schiff's reagent primarily binds DNA, although faint staining of other structures, for example cell walls, may also be observed. This further facilitates the identification of individual cell types. The method is described for maize, but we have also used this technique to successfully observe meiosis in ovules of *Hieracium aurantiacum*, *Arabidopsis thaliana*, and *Eschscholzia californica* (data not shown).

2 Materials

2.1 Plants

We grow plants in the field or at the Biotron in a Conviron BDW120 growth room with lighting provided by 48 metal halide bulbs (Model MS400W/HOR; Venture Lighting International, Inc., Solon, OH, USA) and 48 incandescent bulbs (100 W). Plants are grown under a 16 h light/8 h dark cycle at 25°C during the light and 20°C in the dark. Soil used was as described by Conner et al. (14).

2.2 Fixation Materials

1. Fixation buffer: a 90:2:5 mixture of 50% ethanol, formalin, and acetic acid. The fixative should be made fresh just prior to use and not kept. It is generally useful to make up 200 ml of

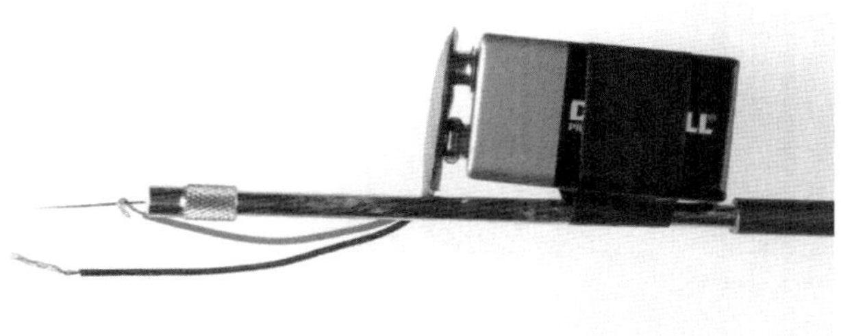

Fig. 1 Tungsten wire holder and 9 V battery

fixation buffer at a time to enable several ears to be fixed. To do this, add 186 ml of 50% ethanol to a 200 ml Schott bottle. In a fume hood add 10 ml of acetic acid and 4 ml of formalin.

2. Ethanol solutions: 95% and 70% ethanol. Make up at least 500 ml each.

3. 50 ml tubes.

4. A sharp pocket knife.

2.3 Staining Solutions

1. 1 M HCl: To make 100 ml of 1 M HCl, add 8.2 ml of 37% HCl to 91.8 ml of distilled water (see Note 1).

2. 5.8 M HCl: To make 100 ml of 5.8 M HCl add 50 ml of 37% HCl to 50 ml of distilled water (see Note 1).

3. Schiff's reagent (Sigma Aldrich, St. Louis, MO, USA).

4. Ethanol series for dehydration: 30%, 50%, 70%, 90%, 95%, and 100% (see Note 2).

5. 50% ethanol: 50% immersion oil (catalog number 1.04699.0100; Merck, Whitehouse Station, NJ, USA).

2.4 Dissection Materials

1. Dissecting microscope with light source, e.g., Leica MZ7.5 dissecting microscope (Leica, Bensheim, Germany).

2. Tungsten wire (catalog number 356972; Sigma Aldrich, St Louis, MO, USA).

3. Needle holders (catalog number L110/01; S Murray & Co, Surry, England).

4. 2 M NaOH.

5. 9 V battery.

6. 9 V battery snap connector clip (Fig. 1).

7. 50 ml tubes.

8. Glass microscope slides.

9. Coverslips: use no. 1.5 thickness, 22×22 mm (see Note 3).

10. Pre-cut labels, e.g., Teeny Tough Tags for 0.2 ml PCR tubes (Diversified Biotech, Boston, MA, USA).

11. Filter paper (see Note 4).

12. Clear nail varnish.

3 Methods

3.1 Fixation

1. Immature ears are harvested prior to silk emergence. Carefully remove the husk leaves from the ear, so that the florets are exposed (see Note 5).

2. Place the ear in a 50 ml tube and cover with an excess of fixative. Leave the ears in fixative at room temperature for 24 h.

3. Carefully rinse the ears several times in 95% ethanol and then store in 70% ethanol at 4°C (see Note 6).

3.2 Staining

1. Rinse ears in distilled water at least three times, with enough water to cover the ear completely. The ears do not need to sit for any length of time in the water, just long enough to be covered.

2. Drain distilled water from tube and add enough 1 M HCl to cover the ear. Leave at room temperature for 1.5 h. Swirl and tap the tube gently to remove any air bubbles that may be trapped near the surface of the tissue.

3. Replace 1 M HCl with 5.8 M HCl. Leave at room temperature for 2 h.

4. Replace 5.8 M HCl with 1 M HCl and incubate for 1 h at room temperature.

5. Rinse the ears at least three times with distilled water. The ears do not need to sit for any length of time in the water, just long enough to be covered.

6. Add enough Schiff's reagent to completely cover the ears and leave at 4°C over night.

7. Dehydrate the ears through an ethanol series: 30%, 50%, 70%, 90%, 95%, and twice in 100% ethanol. Leave the ears for at least 30 min in each ethanol solution.

8. Remove the last ethanol wash and cover the ears with a mixture of 50% immersion oil ethanol and 50% ethanol, and store overnight at 4°C.

3.3 Dissection

1. Prepare slides. Two tough tags are stuck to each slide, approximately 0.5 cm apart. This enables a bridge to be formed that the coverslip sits on to prevent the samples from being squashed. Make sure that the tough tags are completely flattened against the slide to ensure that the cover glass will lie flat.

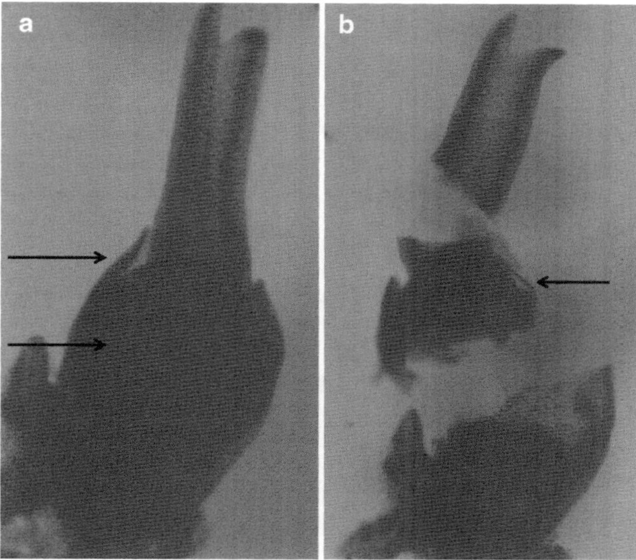

Fig. 2 Dissection of ovules from immature maize florets. (**a**) Arrows indicate where cuts should be made. (**b**) Arrow indicates the section of the floret containing the ovule and meiotic cells

2. Prepare tungsten wire needles. Cut a 3–4 cm strip of tungsten wire, and place in the needle holder. Connect with the snap connector clip to the 9 V battery. Attach the tungsten wire to the positive terminal wire (Fig. 1). Dip the negative wire and the tip of the tungsten wire into the 2 M NaOH solution, until the tungsten wire becomes very sharp. Dissections are easiest with two such tungsten wire needles.

3. Place a drop of immersion oil between the two tough tags on the glass slide, and a floret from the stained ear. Under the dissection scope, carefully cut the silk off (Fig. 2a, top arrow). Slice the floret horizontally approximately half of the way down from the silk attachment site (Fig. 2a, bottom arrow). This middle portion of the floret (Fig. 2b, arrow) will contain the meiotic cells. Dissect away the developing integuments revealing the nucellar tissue containing meiotic cells. Clear away waste floret tissue with filter paper. Add another drop of immersion oil, on the selected dissected material and place a coverslip over the top so that it is sitting on the tough tag bridge. Seal with nail varnish.

3.4 Microscopy

We use either a Leica SP2 or SP5 scanning laser confocal microscope to analyze our samples. Excitation wavelength is 488 nm, and emitted light is detected between 640 and 740 nm. Schiff's reagent is quite a robust stain, and can withstand quite high laser power levels.

We routinely use 40×, 63×, and 100× oil immersion lenses. Because the samples are mounted in oil, we use 8× line averaging in preference to frame averaging to obtain sharp images.

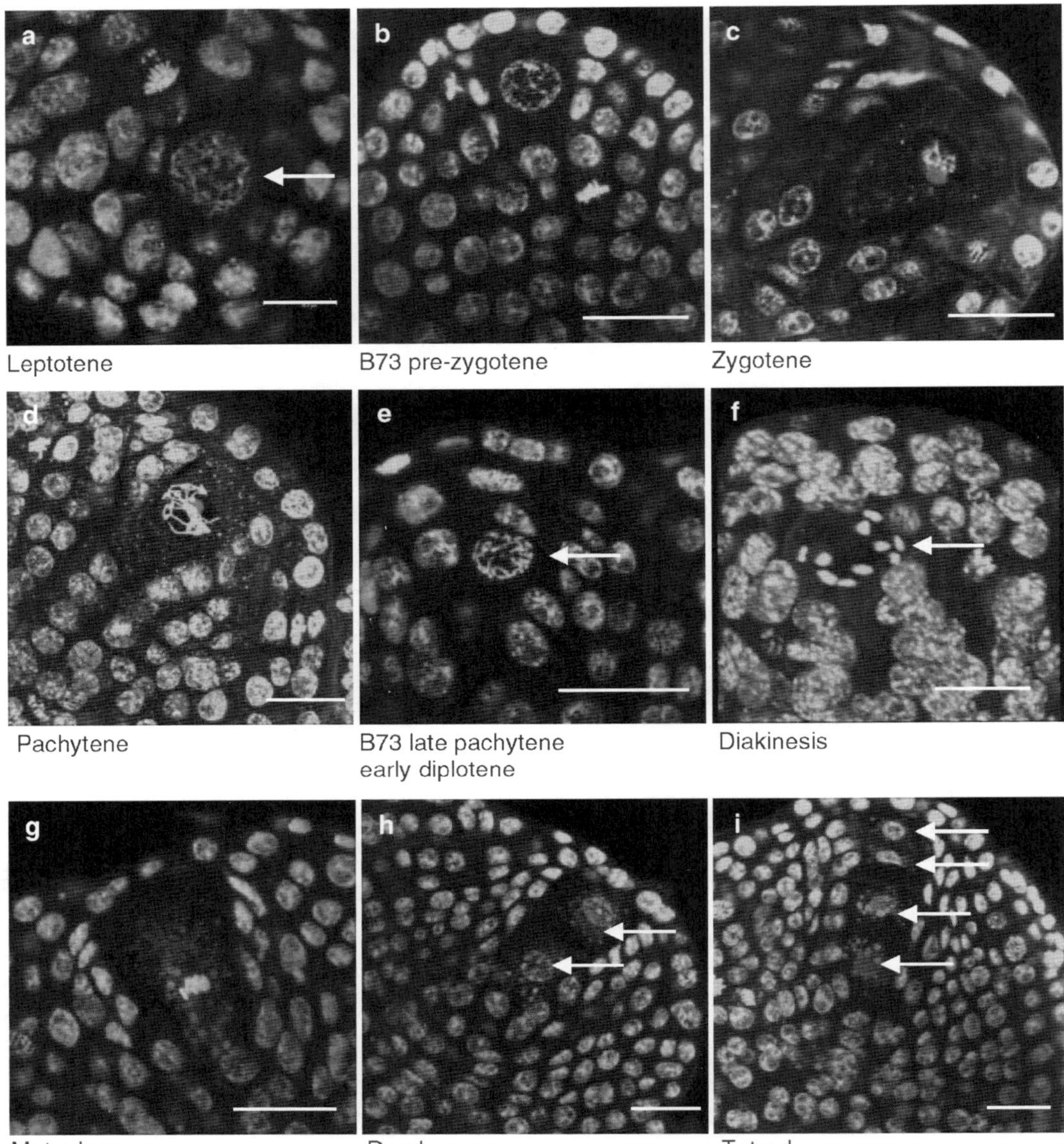

Fig. 3. Confocal analysis of female meiosis in the sequenced inbred line B73. *Arrows* indicate meiotic nuclei. Bars are 20 μm unless otherwise indicated. (**a**) Leptotene. The megaspore mother cell has a rectangle-like form. Depending on the orientation of the dissected ovule relative to the scanning angle of the confocal, this is not always obvious. However, meiotic cells can easily be recognized by their much larger nuclei and sometimes prominent large nucleolus. The nucleus of the megaspore mother cell has occupied a large volume relative to the surrounding ovule tissue. The position of the megaspore mother cell was always one or two cell layers subepidermal. Leptotene chromosomes appear as fine threads. (**b**) Pre-zygotene. (**c**) Zygotene. Telomeres cluster at one end of the nucleus. (**d**) Pachytene. Chromosomes appear as thick threads. (**e**) Late pachytene-early diplotene. Chromosomes further condense. (**f**) Diakinesis. Chromosomes appear very short. Ten bivalents can be counted. (**g**) Metaphase. Chromosomes align and group together. (**h**) Dyad. Two cells are formed, the products of meiosis I. (**i**) Tetrad. The dyad divides without further chromosome replication yielding four haploid spores. *Arrows* indicate the four nuclei of the tetrad

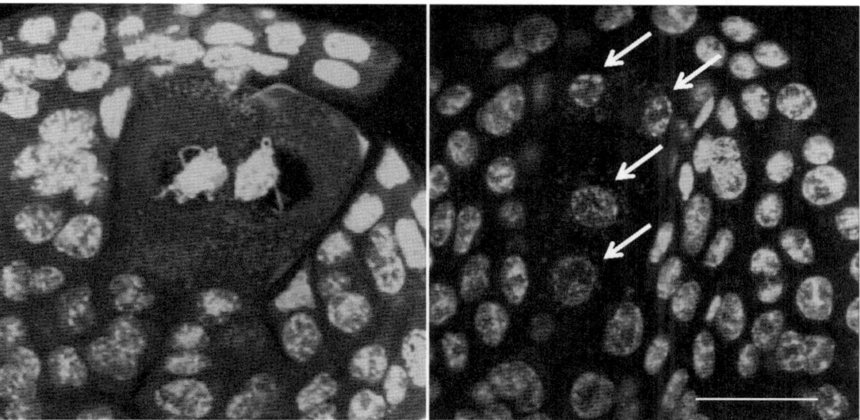

Fig. 4 Aberrations in meiosis. (**a**) Two megaspore mother cells appear instead of one. (**b**) Tetrad stage. In most cases the tetrad is formed by horizontally aligned divisions of the dyad. In this case the upper dyad divided at right angles to the lower dyad, and the resulting four tetrad nuclei are not aligned (*arrows*)

Optical sections are recorded and assembled into 3D images. The images presented here were collected with the software operating the Leica SP5 confocal microscope, and have not undergone any secondary deconvolution processing.

It is possible to quantify signal levels and determine ploidy for individual nuclei using Imaris software (Bitplane, Zürich, Switzerland) (15). The images presented here (Figs. 3 and 4) were collected with the software operating the Leica SP5 confocal microscope and have not undergone any secondary deconvolution processing.

4   Notes

1. Always add concentrated acid to water, not the other way round.

2. We use technical grade ethanol; there is no need for ultrapure ethanol.

3. Most objectives are designed for 1.5 thickness (0.17 mm). The use of coverslips outside thickness 1.5 may result in a reduction in optical resolution.

4. Whatman filter paper torn into small pieces, approximately 1–2 cm square, works best. The material to be discarded is more easily picked up by the torn edge.

5. It is generally useful to harvest secondary ears where possible, to allow primary ears to be pollinated and lines to be maintained or used for other genetic or molecular analyses.

6. Ears can be kept for several weeks at 4°C at this stage.

Acknowledgments

Thanks to the Lincoln University Biotron Facility (New Zealand) for access to plant growth room facilities. This work was supported by New Zealand's Ministry of Science and Innovation (contract C02X0808) and the University of Zürich, Switzerland.

References

1. Drews GN, Wang DF, Steffen JG, Schumaker KS, Yadegari R (2011) Identification of genes expressed in the angiosperm female gametophyte. J Exp Bot 62:1593–1599

2. Evans MMS, Grossniklaus U (2009) The maize megagametophyte. In: Bennetzen JL, Hake S (eds) Handbook of maize: its biology. Springer, New York, pp 79–104

3. Ma H, Sundaresan V (2010) Development of flowering plant gametophytes. Curr Top Dev Biol 91:379–412

4. Sprunck S, Gross-Hardt R (2011) Nuclear behavior, cell polarity, and cell specification in the female gametophyte. Sex Plant Reprod 24:123–136

5. Bhatt AM, Canales C, Dickinson HG (2001) Plant meiosis: the means to 1 N. Trends Plant Sci 6:114–121

6. Dawe RK (1998) Meiotic chromosome organization and segregation in plants. Annu Rev Plant Physiol Plant Mol Biol 49:371–395

7. Hamant O, Ma H, Cande WZ (2006) Genetics of meiotic prophase I in plants. Annu Rev Plant Biol 57:267–302

8. Armstrong SJ, Jones GH (2003) Meiotic cytology and chromosome behaviour in wild-type *Arabidopsis thaliana*. J Exp Bot 54:1–10

9. Zickler D, Kleckner N (1999) Meiotic chromosomes: integrating structure and function. Annu Rev Genet 33:603–754

10. Golubovskaya I, Avalkina N (1994) Protocol for preparing maize macrospore mother cells for the study of female meiosis and embryo-sac development. In: Freeling M, Walbot V (eds) The maize handbook. Springer, New York, pp 450–456

11. Grimanelli D, Garcia M, Kaszas E, Perotti E, Leblanc O (2003) Heterochronic expression of sexual reproductive programs during apomictic development in tripsacum. Genetics 165:1521–1531

12. Kiesselbach TA (1949) The structure and reproduction of corn. Cold Spring Harbor Laboratory Press, Cold Spring Harbor, NY

13. Braselton JP, Wilkinson MJ, Clulow SA (1996) Feulgen staining of intact plant tissues for confocal microscopy. Biotech Histochem 71:84–87

14. Conner AJ, Williams MK, Abernethy DJ, Fletcher PJ, Genet RA (1994) Field performance of transgenic potatoes. New Zeal J Crop Hort 22:361–371

15. Barrell PJ, Grossniklaus U (2005) Confocal microscopy of whole ovules for analysis of reproductive development: the *elongate1* mutant affects meiosis II. Plant J 43:309–320

Chapter 6

Three-Dimensional Acrylamide Fluorescence In Situ Hybridization for Plant Cells

Elizabeth S. Howe, Shaun P. Murphy, and Hank W. Bass

Abstract

Plant meiosis involves complex and dynamic processes that occur within the space inside the nucleus. Direct inspection of meiotic chromosomes by fluorescence microscopy has been used to investigate many of these processes. In particular, optical sectioning microscopy of fluorescence in situ hybridization (FISH)-stained nuclei provides three-dimensional spatial information about the organization and distribution of specific sequences and chromosomal loci within the nucleus. Here we provide a fully detailed three-dimensional (3D) acrylamide FISH method for the analysis of plant meiotic nuclei. Several examples illustrate the versatility of this technique for the investigation of meiotic telomere dynamics in maize, *Arabidopsis*, and oat. Additional examples of 3D FISH include chromosome painting in a maize chromosome-addition line of oat and telomere FISH with maize nuclei from plants expressing a fluorescently tagged fusion protein, histone H2B-mCherry.

Keywords Molecular cytology, Meiosis, Nucleus, 3D microscopy

1 Introduction

Preservation of chromatin structure for cytological analysis by means of "Buffer A" was demonstrated for *Drosophila* polytene chromosomes by J.W. Sedat and members of his laboratory (1, 2). His laboratory also showed that acrylamide provides an optically clear, heat-stable gel matrix suitable for cytology, allowing for immobilization and manipulation of fixed cells for cell staining and imaging (3). The three-dimensional (3D) acrylamide FISH procedure developed by H.W. Bass in 1997 (4) combined the Buffer A-based formaldehyde fixation with acrylamide embedding. The resulting 3D molecular cytology afforded clear views of maize meiotic chromosomes and telomeres (4–6). The ability to visualize the spatial distribution of FISH signals in the nucleus in three dimensions is particularly important for the study of meiosis, during which homologs pair, telomeres are redistributed, and chromatin

Wojciech P. Pawlowski et al. (eds.), *Plant Meiosis: Methods and Protocols*, Methods in Molecular Biology, vol. 990, DOI 10.1007/978-1-62703-333-6_6, © Springer Science+Business Media New York 2013

structure and nuclear architecture undergo dramatic changes during the progression of meiotic prophase (7, 8).

Since its development, this 3D acrylamide FISH method has been widely used to characterize the distribution of meiotic chromosomes, specific loci, and telomeres in plants (5, 9–18). This robust and versatile technique has been modified for examination of viral DNA replication in tobacco leaf cells (19), as well as analysis of allopolyploid subgenome distribution in *Torenia* (20). In addition to FISH, the acrylamide embedding procedure is also compatible with immunocytochemistry (9, 10, 14, 16, 18, 21) and the visualization of endogenous fluorescent proteins, as described below for histone H2B-mCherry.

2   Materials

2.1   Plants

Plants used to develop this protocol were grown in greenhouses or growth chambers. They included a maize line expressing a histone H2B-mCherry fusion protein (event C01; Figs. 1 and 3; described in detail by Howe et al. (22)), a maize inbred line (A344+; Fig. 2), a maize-9 chromosome addition line of oat (om9; Fig. 4), and wild-type *Arabidopsis* (courtesy of H. Cui, ecotype Columbia; Fig. 5).

2.2   Fixation

1. 10× Meiocyte Buffer A salts (10× MBA–––): 800 mM KCl, 200 mM NaCl, 150 mM PIPES, 20 mM EGTA, 5 mM EDTA, 1 M NaOH. Combine ingredients from solid, high-quality stocks, adjust pH to 6.8 with NaOH, bring to final volume, and sterilize by filtration (through a 0.2–0.4-μm filter). Aseptically

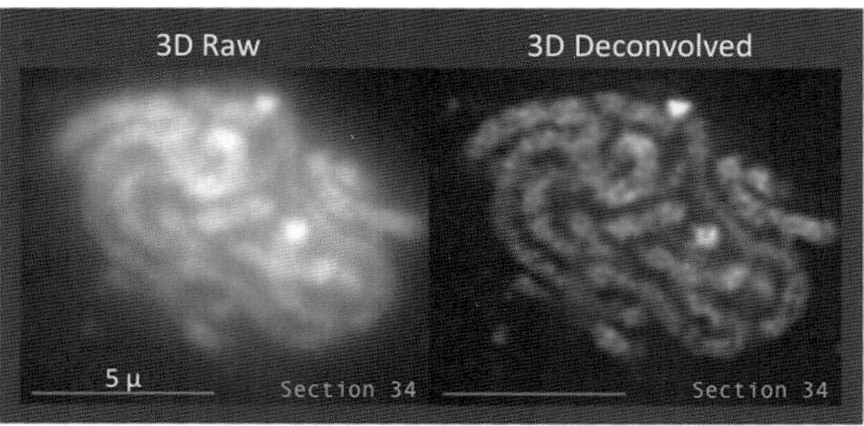

Fig. 1 Deblurring by 3D deconvolution. Three-dimensional (3D) telomere FISH (not shown) was carried out on acrylamide-embedded cells from meiotic-stage anthers. The original data (70 optical sections) were computationally cropped in all directions (*X*, *Y*, *Z*). The 4′, 6-diamino-2-phenylindole (DAPI) image of optical section 34 from a single pachytene nucleus is shown before (3D Raw) and after (3D Deconvolved) deconvolution. The deconvolved image reveals the preservation of morphological detail in the chromosome fibers, even after 3D acrylamide FISH

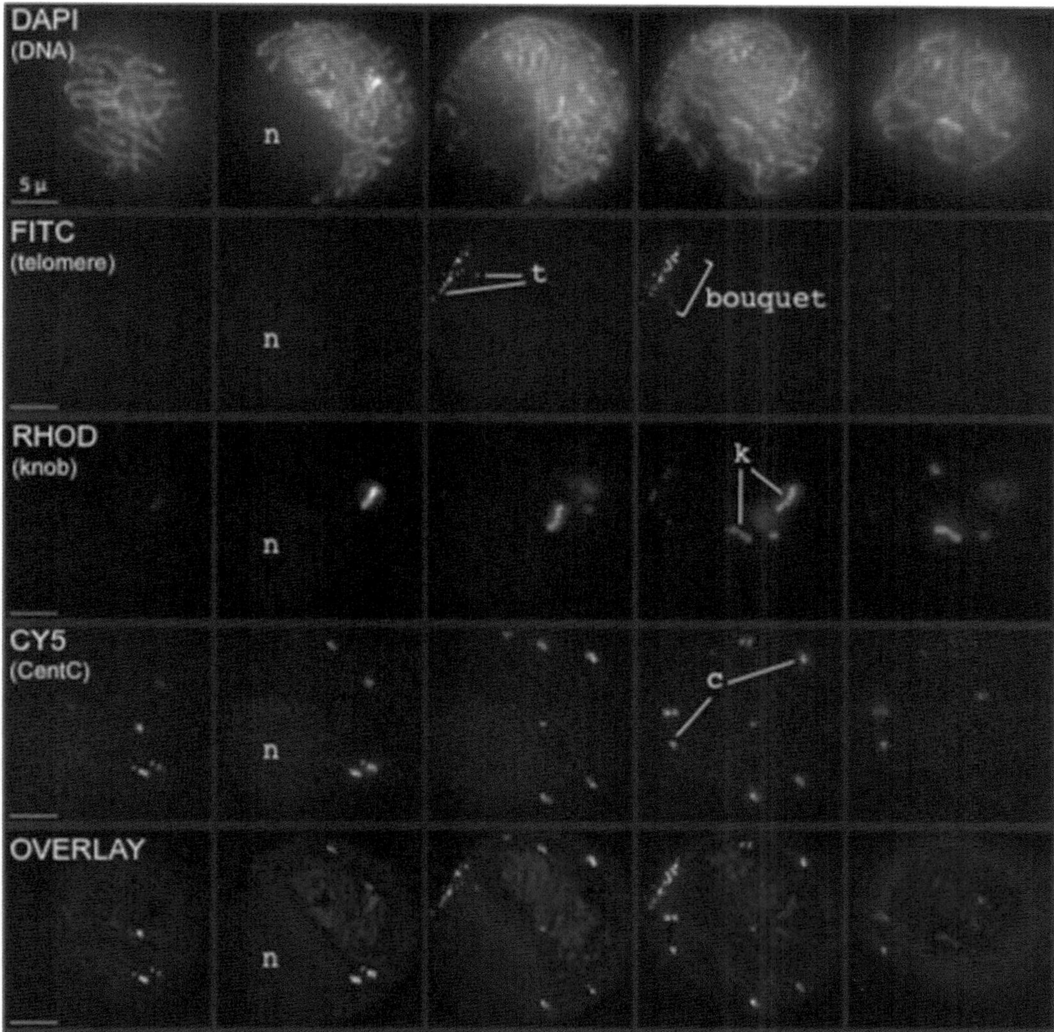

Fig. 2 3D FISH of a bouquet-stage maize nucleus. 3D acrylamide FISH was carried out with three different fluorescent probes, counter-stained with DAPI, and imaged at all four wavelengths. The 3D data (70 optical sections) were converted to five sequential projections, each with an effective focal plane of one-fifth the nuclear depth. This display conveys all the data from a single 3D set as five sequential projections. The wavelenths/fluorophores (DAPI, FITC, RHOD/rhodamine, and CY5) are indicated, along with the structures they stain (*in parentheses*, DNA, telomere, knob, and centromere). The location of the nucleolus (n), telomere FISH signals (t), telomere cluster bouquet (bouquet, *white bracket*), knob FISH signals (k), and centromere FISH signals (c) are indicated

dispense into new, sterile 50-ml polypropylene conical tubes. Store at –80°C for the long term or at –20°C for use within 6 months.

2. 1.6 M D-sorbitol. Sterilize by filtration and dispense aliquots into sterile 50-ml polypropylene conical tubes. Store at –80°C.

3. 1,000× polyamines (PA): 0.15 M spermine (spermine tetra HCl; catalog no. S-2876; Sigma Aldrich, St. Louis, MO, USA), 0.5 M spermidine (catalog no. AAA-19896-06; VWR, Wayne, PA, USA). Prepare freshly, sterilize by filtration into 1.5-ml polypropylene microcentrifuge tubes, and store at −80°C in aliquots of a size for one-time thawing and use.

4. 1,000× dithiothreitol (DTT): 1.0 M DTT, 0.01 M NaOAc. Prepare freshly on ice, sterilize by filtration, and dispense aliquots into 1.5-ml polypropylene microcentrifuge tubes. Store at −80°C.

5. 1× Meiocyte Buffer A (1× MBA+++): 1× Buffer A salts, 0.32 M sorbitol, 1× DTT, 1× polyamines. Prepare buffer aseptically from frozen stocks just before use. Store at 4°C.

6. 1× MBA+++ plus 4% formaldehyde (MBA-Fix): 1 MBA+++, 4% paraformaldehyde (catalog no. 04018; Polysciences, Warrington, PA, USA). Prepare freshly in a sterile 50 ml polypropylene conical tube. Add formaldehyde just before use on plant material. Vortex well and store on ice.

7. 24-well 3.4-ml cell-culture cluster plates (Costar; catalog no. 3524; Fisher Scientific, Suwanee, GA, USA).

8. Microscalpel (5 mm Depth 15° tab; catalog no. PE3015-5; Oasis, Glendora, CA, USA).

9. Fine forceps (Dumont #5; catalog no. 11252-30; Fine Sciences Tools, Foster City, CA, USA).

2.3 Acrylamide

1. Bovine serum albumin (BSA): 200 mg/ml BSA, 1× PBS. Store at −80°C. BSA (IgG-free, protease-free BSA; catalog no. 001-000-16; Jackson Immuno Research, West Grove, PA, USA). Whatman-filter and dispense aliquots into 1.5-ml polypropylene microcentrifuge tubes.

2. 30% acrylamide stock: 30:3.3 (% vol:vol) mixture of acrylamide (catalog no. BP 170-100; Fisher Scientific, Suwanee, GA, USA) and bisacrylamide (catalog no. BP 171-25; Fisher Scientific, Suwanee, GA, USA). Prepare fresh, sterilize by filtration, and store in an amber glass container at 4°C (see Note 1).

3. 3× acrylamide mix: 15% acrylamide stock, 1× Meiocyte Buffer A+++, 1× PA, 1× DTT, 0.32 M sorbitol. Prepare freshly in a sterile 15-ml polypropylene conical tube; vortex well. Dispense aliquots into 1.5-ml polypropylene microcentrifuge tubes. Degas solution for 10 min in a SpeedVac at room temperature with the tube lids open or holes in tops of tubes (see Note 2). Remove from SpeedVac and leave at room temperature with lids closed until use.

4. 20% ammonium persulfate in ddH$_2$O. Prepare freshly in 1.5-ml polypropylene microcentrifuge tubes. Vortex thoroughly to dissolve (see Note 3).

5. 20% sodium sulfite (anhydrous) in ddH$_2$O. Prepare freshly in 1.5-ml polypropylene microcentrifuge tubes. Vortex thoroughly to dissolve.

2.4 FISH

1. 20× Standard Saline Citrate (20× SSC): 3.0 M NaCl, 0.3 M Na$_3$Citrate. Sterilize by filtration, autoclave, and store at room temperature.

2. 10× Phosphate Buffered Saline (10× PBS): 1.37 M NaCl, 27 mM KCl, 100 mM dibasic Na$_2$HPO$_4$·7H$_2$O, 20 mM monobasic KH$_2$PO$_4$. Adjust pH to 7.4 with HCl, sterilize by filtration, autoclave, and store at room temperature.

3. 4′, 6-diamino-2-phenylindole (DAPI): 3 µg/ml (catalog no. D9542; Sigma Aldrich, St. Louis, MO, USA).

4. Prehybridization Buffer: 50% formamide (catalog no. EM-4610; VWR, Wayne, PA, USA), 2× SSC. Prepare buffer freshly in a sterile 15-ml polypropylene conical tube (see Note 4).

5. DNA probes for FISH (see Note 5).

6. Hybridization Buffer: 50% formamide, 2× SSC, probe(s). Prepare buffer freshly in a sterile 15-ml polypropylene conical tube. Dilute concentrated probe in Hybridization Buffer to 3 µg/ml (see Note 6).

7. Wash 1: 4× SSC. This and the other numbered washes are used in Subheading 3.7.

8. Wash 2: 2× SSC, 2× PBS.

9. Wash 3: 1× PBS, 0.2% Tween-20 (from 1% sterile stock).

10. Wash 4: 1× PBS, 1 mM DTT.

11. Wash 5: 1× PBS.

12. Wash 6: 1× PBS, 3.0 µg/ml DAPI.

13. Wash 7: 1× PBS, 1 mM DTT.

14. Glass coverslips (Thomas Scientific Red Label 22x30 #1.5; catalog no. 6663K19; VWR, Wayne, PA, USA) (see Note 7).

15. Rubber cement (Elmer's).

16. 9.0-×9.0-cm square petri plates (catalog no. 08-757-11A; Fisher Scientific, Suwanee, GA, USA).

17. Vectashield (catalog no. H-1000; Vector Laboratories, Burlingame, CA, USA).

18. Clear nail polish (Sally Hanson Clear Hard as Nails).

19. Glass slides (Gold Seal; catalog no. 3010; VWR, Wayne, PA, USA).

20. Epifluorescence microscope (e.g., Olympus IX-70 wide-field microscope, equipped with oil-immersion lens (60x NA 1.4 PlanApo, Olympus), cooled CCD camera, and DeltaVision software (SoftWorx; Applied Precision, Issaquah, WA, USA)).

3   Methods

3.1 Fixation of Plant Material

1. Microdissect anthers from freshly harvested intact florets and float on or submerge in MBA-Fix in 12- or 24-well culture plates. Fix anthers by rotary shaking at room temperature for 1 h. For fixation and all subsequent washes, use 2 ml for every 10–20 anthers (see Note 8).

2. After fixation, gently transfer anthers with fine forceps to 1× MBA+++ in adjacent wells. Wash by rotary shaking at room temperature for 30 min per wash, 2–3 washes total.

3. Store in 1× MBA+++ at 4°C or use immediately (see Note 9).

3.2 Preparation of Slides and Coverslips

1. Label slide corner with a diamond marker.

2. Wash plain glass slides with 70% ethanol and Kimwipes, or other nonabrasive, low-linting, low-extractables paper tissues, until they are residue-free, and remove dust with lens paper or filtered air. Hold clean slides in square petri plates (three per plate) to keep free of dust.

3. Paint a ring of nail polish in the middle of each slide. This process usually requires two large drops of nail polish. The ring should have a ~1- to 2-cm^2 open area for the acrylamide pad (see Note 10).

4. Let slides dry at room temperature for at least 3 h or until the nail polish no longer smells of acetone and hold in petri plates to keep free of dust.

5. Prepare $22 \times 30 \times 1.5$ coverslips by rubbing them clean with dry lens paper to remove all visible residue.

3.3 Extrusion of Fixed Meiocytes

1. For microdissection of anthers, place a 50- to 200-μl drop of 1× MBA on a circle of parafilm in the inverted lid of a standard plastic petri plate.

2. Add five to eight fixed anthers to the droplet and push them completely under the droplet before cutting them open to release or extrude the meiocytes.

3. To extrude the meiocytes, hold the base of the anther with forceps and use a microscalpel to make a transverse incision across the end of the anther.

4. Hold the base of the anther with the forceps and gently tap or squeeze out the meiocyte column with the blunt side of the scalpel. Start tapping the blade of the scalpel at the base of the anther and slowly move it toward the cut end (see Note 9).

5. Once all the anthers for that slide have been dissected, discard the large non-meiocyte tissue clumps before transferring the cells to the slide for embedding in polyacrylamide.

3.4 Embedding of Meiocytes in 5% Polyacrylamide, 1× MBA+++

1. Transfer 13 µl of meiocytes from the petri plate to the slide, spreading them inside the nail-polish ring (see Note 11).

2. Rapidly combine 25 µl of 20% APS and 25 µl of 20% sodium sulfite with 0.5 ml of 3× acrylamide mix. Mix quickly by vortexing briefly and add 6.5 µl of this activated mix to the drop of meiocytes on the slide, stirring gently with the pipet tip. Acrylamide will polymerize in the tube in 20–30 s but much more slowly on the slide.

3. Cover with a cleaned coverslip, avoiding bubbles inside the nail polish ring. Excess buffer may spill over and can be removed later by aspiration, after the acrylamide gel pad has formed.

4. Allow acrylamide to polymerize under the coverslip for 20 min.

5. Remove the coverslip by slowing prying it up with razor blade, keeping the acrylamide on the slide (see Note 12) and gently cover the acrylamide pad with 200 µl of 1× MBA+++. The buffers should sit atop the acrylamide pad but stay within the nail polish ring during all subsequent buffer exhange/wash steps.

6. Wash the acrylamide pad three times with 1× MBA+++ (see Note 13).

3.5 Prehybridization

1. Aspirate off all of the remaining solution from the slides and wipe the slides with lens paper to remove all visible residues outside of the nail-polish ring.

2. Equilibrate the pad with Prehybridization Buffer by three additions of buffer, as described in Subheading 3.4, step 6.

3.6 Hybridization

1. Clean the slides as described in Subheading 3.5, step 1.

2. Add 60 µl of Hybridization Buffer to the pad. Use the pipet tip to distribute the buffer as evenly as possible over the acrylamide pad.

3. Remove the Hybridization Buffer by tilting the slide and aspirating the droplet from the bottom. Add 60 µl of probe mix to the center of a 22×30×1.5 coverslip on lens paper. Invert the slide and touch it to the drop to mount the coverslip on the slide, avoiding air bubbles as much as possible.

4. Wipe any residue from around the edges of the slide and seal the coverslip by addition of an excessive amount, up to 1 ml, of rubber cement around the edges.

5. Place slides on a slide warmer or heat block or in an incubator at 37°C. Allow rubber cement to dry for at least 15 min. Slides can be left at 37°C until all of the other slides are prepared.

6. Heat denature the probe and target sequences in the nuclei by placing the slides on a heat block at 92°C for 6 min (see Note 14).

7. Place slides directly in prewarmed containers and incubate at 37°C overnight.

<table>
<tr><td>3.7 Washing
and Mounting</td><td>

1. Remove slides from the 37°C incubator and roll the rubber cement off gently with gloved hands, being careful not to pull or distort the acrylamide pad under the coverslip (see Note 15).

2. Remove the coverslip by inserting a razor blade under one corner and prying it off gently and very slowly. Hold the slide at an angle so that the coverslip will fall on the lab bench and not back onto the acrylamide pad in case it slips off the razor during removal.

3. Wash and DAPI stain using the same buffer-exchange procedures as described in Subheading 3.4, step 6.

4. Wash five times with Wash 1, 5 min each.

5. Wash three times with Wash 2, 5 min each.

6. Wash three times with Wash 3, 5 min each.

7. Wash three times with Wash 4, 5 min each.

8. Wash three times with Wash 5, 5 min each.

9. Wash twice with Wash 6, 5 min for the first wash, 20 min for the second wash.

10. Wash five times with Wash 7, 5 min each.

11. Clean the slides with lens paper, using several sheets per slide, to remove all buffers around the acrylamide pad so that the nail polish will make a good seal. Be careful not to touch the acrylamide pad itself.

12. Mount in Vectashield by adding 100 µl Vectashield and rotating the slide by hand to mix and distribute the thick mounting medium uniformly over the pad. Allow for buffer exchange by rotary shaking for 10 min at higher speed than used for aqueous buffers. Remove and reapply twice more, cleaning the slide again after the last removal of Vectashield.

13. Clean coverslip with lens paper and place it on a clean piece of lens paper at the edge of the bench.

14. Remove any remaining nail polish ring with forceps and wipe clean to remove any visible debris.

15. Mount by placing 20–30 µl Vectashield on the cleaned coverslip. Invert the slide and mount the coverslip upside down, making contact at an angle to get uniform bubble-free spread. If wrinkles in the pad appear, attempt to flatten them with gentle pressure or remounting.

16. Seal the coverslip with nail polish.

17. Let the nail polish air dry completely before microscopy. Image immediately for best results or store in air-tight, light-tight containers at –20°C for later imaging.

</td></tr>
</table>

3.8 Three-Dimensional Image Collection

Image data can be collected with any epifluorescence microscope workstation. Deconvolution and confocal microscopes are particularly useful for collecting 3D data, but conventional epifluorescence microscopes can also be used to collect optical-section images from samples in acrylamide. For the examples shown here, we collected and deconvolved images using a DeltaVision deconvolution microscope workstation (23, 24).

For data collection, each optical section can be imaged with up to four different filter sets for the selective detection of DAPI, FITC, rhodamine, and/or Cy5. Typically, 60–100 optical sections (at 0.2-μm steps) are collected, producing a multiple-wavelength set of 3D data. For these data, voxel dimensions are typically $0.07 \times 0.07 \times 0.2$ μm. Figure 1 shows the effect of deblurring by iterative 3D deconvolution. Figures 2, 3, 4, 5 show different examples of 3D acrylamide FISH in several species, illustrating the power of this technique to resolve the spatial arrangement of specific DNA sequences within individual nuclei.

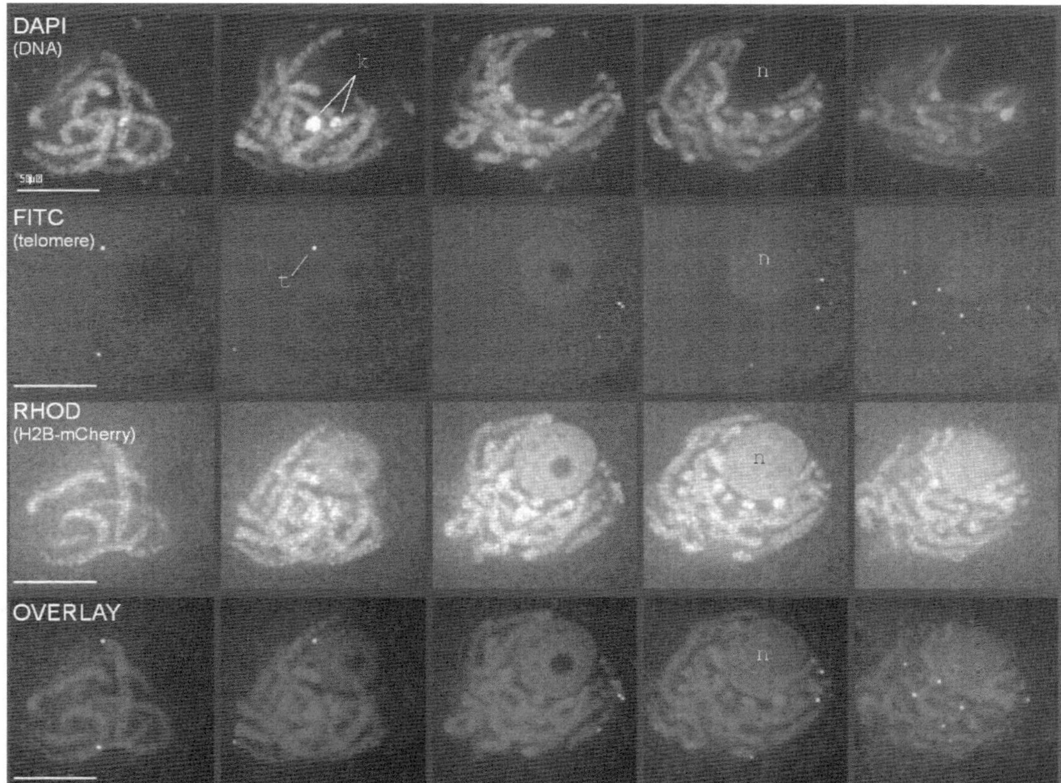

Fig. 3 3D acrylamide FISH showing telomeres and fluorescently tagged Histone H2B-mCherry. Staining, data collection, and image display were carried out as described for Fig. 2. The locations of the nucleolus (n), telomere FISH signals (t), and the knobs (k) are indicated

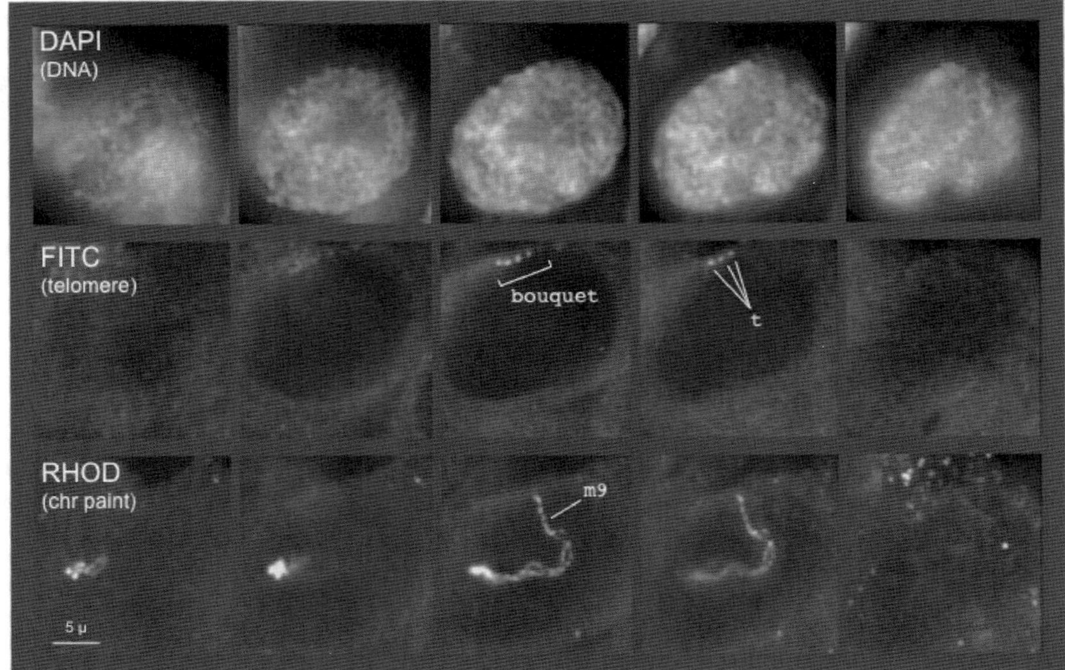

Fig. 4 3D acrylamide FISH in oat. 3D acrylamide FISH showing whole-chromosome painting and the telomere bouquet in an oat addition line containing maize chromosome 9 (5). Staining, data collection, and image display were carried out as described for Fig. 2. The locations of the telomere cluster bouquet (bouquet), telomere FISH signals (t), and chromosome paint (m9) are indicated

3.9 Display of Image Data

The 3D data sets were cropped around individual nuclei. Sequential maximum-intensity projections are shown in Figs. 2, 3, 4, 5. These gray-scale and color overlay images were adjusted for brightness and contrast with linear scaling.

4   Notes

1. Acrylamide is a potent neurotoxin. Avoid contact with the powder by using a fume hood. Dissolved acrylamide is not volatile, but avoid any direct contact by following routine procedures for hazardous materials.

2. Degassing increases the rate and efficiency of acrylamide polymerization.

3. Prepare the APS and sodium sulfate stocks just before use and use them within 30 min.

4. For eight slides, prepare 10 ml of Prehybridization Buffer.

5. Probes vary in size, complexity, and synthesis. They should be optimized for each type of experiment. For abundant tandemly

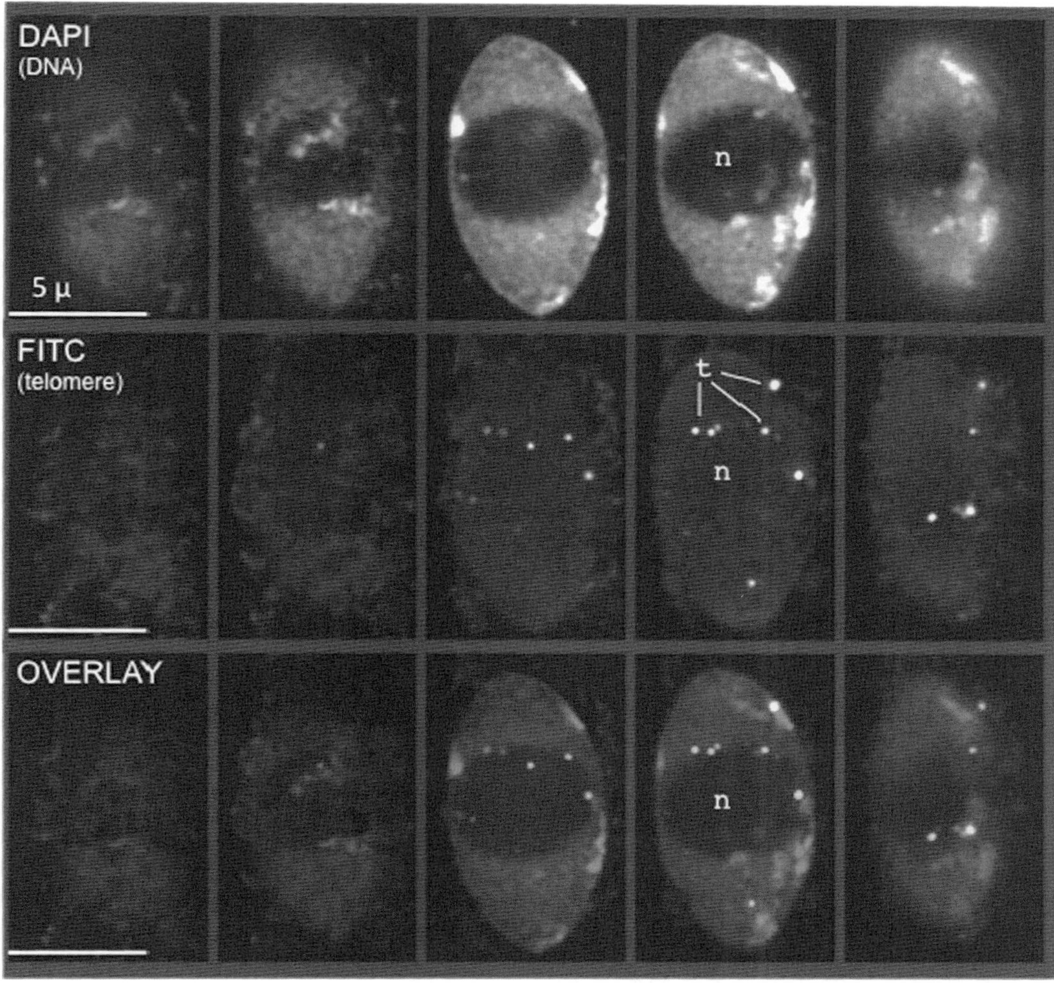

Fig. 5 3D acrylamide FISH in *Arabidopsis*. 3D acrylamide FISH showing telomere distributions in an early prophase nucleus of *Arabidopsis*. Staining, data collection, and image display were carried out as described for Fig. 2. The locations of the nucleolus (n) and telomere signals (t) are indicated. Telomere colocalization with the nucleolar periphery in pollen mother cells is typical for *Arabidopsis* and evident in this example

repeated sequences, such as telomeres, centromeres, knobs, or 5S rDNA, fluorescently labeled oligonucleotide probes are preferred. For maize meiotic nuclei, oligonucleotide probes should be 20–30 nt long and used at a final concentration of 1–3 μg/ml. The telomere FISH probe, for example, is 5′-FITC-CCCTAAACCCTAAACCCTAAACCCTAAA-3′. Dyes used successfully for this purpose include FITC, 6-FAM, ROX, TAMRA, and Cy-5. In addition, most of the highly stable Alexa-Fluor series dyes (Invitrogen, Carlsbad, CA, USA) are available for custom oligonucleotide labeling. Directly labeled oligonucleotide probes are relatively inexpensive, uniformly labeled with exactly one fluorophore per molecule, and easily stored for years in small frozen aliquots. Alternatively, probes can be prepared according to conventional labeling techniques with fluorescently labeled nucleotides, digoxigenin-labeled nucleotides, or biotin-labeled nucleotides. For indirect labeling, standard detection reagents are used at manufacturer's specifications.

6. Fluorescent reagents are subject to photobleaching and should be stored in amber or foil-covered containers for all relevant steps throughout the entire procedure.

7. The use of coverslips with 1.5 thickness is required to match the refractive index of the immersion oil and mounting medium as specified for the API DeltaVision deconvolution microscope. Other microscopy systems may require different coverslip thickness.

8. Proper fixation steps for oat are described by Bass et al. (5). For *Arabidopsis*, flower buds should be fixed under house vacuum and used as source for anthers.

9. We have noticed that extrusion of the meiotic cells from anthers with the tips cut off becomes easier if the anthers are first stored at 4°C for at least 2 weeks. Nuclear morphology and suitability for FISH are well maintained at 4°C if storage is aseptic. Anthers can be used up to 1 year after fixation but are best if used within 2 months after fixation.

10. The nail-polish ring provides a raised border to contain the acrylamide pad and holds the coverslip parallel and close to the slide during the acrylamide-polymerization stage. The "ring" can be a circle, square, or rectangle, but it should be smaller than the coverslip (22×22 or 22×30). The ring typically leaves about a 1-cm^2 area for the acrylamide pad. Let the ring dry thoroughly, typically at least 3 h with aeration or overnight. Extra buffer can spill over the ring when the coverslip is placed over the sample for acrylamide gel polymerization, so the exact size of the enclosed space is not critical.

11. To transfer cells gently with a micropipette and to reduce the tendency of cells to stick to the pipet-tip walls, cut the micropipette tip 2 mm from the end to widen the bore and prepipet MBA+++ buffer supplemented with 20–50 μg/ml BSA in and out of the tip before pipetting the 13 μl of cells from the droplet onto the slide.

12. The low percentage of acrylamide used in this technique is important for efficient buffer exchange but can occasionally result in failure to polymerize. Remove the coverslip by inserting a razor blade under one corner and gently pry it up. Acrylamide at the edge may be unpolymerized, but the acrylamide pad may

already have formed inside the nail-polish ring. The pad can stick to either the slide (preferred) or the coverslip. If it sticks to the coverslip, prepare a new cleaned and labeled slide. Add a small drop of water to the slide and place the coverslip, pad side up, on the clean slide. With a pipet tip, gently press down on the coverslip and remove excess water by aspiration. Avoid getting bubbles under the coverslip.

13. To exchange buffers and wash the cells, place large drops (~200 μl) on top of the acrylamide pad and subject the slide to gentle rotary shaking at room temperature for 5 min. Remove the buffer by aspiration using a pipette tip and house vacuum, being careful not to aspirate the acrylamide pad itself. Tilting the slide to cause the droplet to move off the acrylamide before aspiration reduces the risk of aspirating the acrylamide pad.

14. The duration of heat exposure and denaturation temperature can be modified for optimization. These parameters affect chromosome morphology and signal intensity, and optimal results are a balance of these two factors. To optimize, try one temperature for 4, 6, 8, and 12 min, or try several temperatures for a fixed duration (e.g., 6 min) and compare the resulting images.

15. For the posthybridization washes, the slides will be exposed to a lot of benchtop manipulations and should be covered with foil whenever practical during all of these steps.

Acknowledgments

We would like to thank A.B. Thistle and D.L. Vera for critical reading of the manuscript and insightful comments and T.E. Clemente for providing the histone H2B-mCherry line of maize. This work was supported by a Women in Science Math and Engineering Fellowship to ESH, an American Heart Association predoctoral fellowship to SPM (AHA, Greater Southeast Affiliate, number 0715487B), a CRC-planning grant to HWB (2008), and a National Science Foundation grant to HWB (NSF IOS-1025954).

References

1. Belmont AS, Sedat JW, Agard DA (1987) A three-dimensional approach to mitotic chromosome structure: evidence for a complex hierarchical organization. J Cell Biol 105:77–92

2. Sedat J, Manuelidis L (1978) A direct approach to the structure of eukaryotic chromosomes. Cold Spring Harb Symp Quant Biol 42:331–350

3. Urata Y, Parmelee SJ, Agard DA, Sedat JW (1995) A three-dimensional structural dissection of *Drosophila* polytene chromosomes. J Cell Biol 131:279–295

4. Bass HW, Marshall WF, Sedat JW, Agard DA, Cande WZ (1997) Telomeres cluster *de novo* before the initiation of synapsis: a three-dimensional spatial analysis of telomere

positions before and during meiotic prophase. J Cell Biol 137:5–18

5. Bass HW, Riera-Lizarazu O, Ananiev EV, Bordoli SJ, Rines HW, Phillips RL et al (2000) Evidence for the coincident initiation of homolog pairing and synapsis during the telomere-clustering (bouquet) stage of meiotic prophase. J Cell Sci 113:1033–1042

6. Bass HW, Bordoli SJ, Foss EM (2003) The *desynaptic* (*dy*) and *desynaptic1* (*dsy1*) mutations in maize (*Zea mays* L.) cause distinct telomere-misplacement phenotypes during meiotic prophase. J Exp Bot 54:39–46

7. John B (1990) Meiosis. Cambridge University Press, Cambridge, UK

8. Zickler D, Kleckner N (1999) Meiotic chromosomes: integrating structure and function. Annu Rev Genet 33:603–754

9. Franklin AE, McElver J, Sunjevaric I, Rothstein R, Bowen B, Cande WZ (1999) Three-dimensional microscopy of the rad51 recombination protein during meiotic prophase. Plant Cell 11:809–824

10. Franklin AE, Golubovskaya IN, Bass HW, Cande WZ (2003) Improper chromosome synapsis is associated with elongated RAD51 structures in the maize *desynaptic2* mutant. Chromosoma 112:17–25

11. Cowan CR, Carlton PM, Cande WZ (2002) Reorganization and polarization of the meiotic bouquet-stage cell can be uncoupled from telomere clustering. J Cell Sci 115:3757–3766

12. Golubovskaya IN, Harper LC, Pawlowski WP, Schichnes D, Cande WZ (2002) The *pam1* gene is required for meiotic bouquet formation and efficient homologous synapsis in maize (*Zea mays* L.). Genetics 162:1979–1993

13. Golubovskaya IN, Hamant O, Timofejeva L, Wang CJ, Braun D, Meeley R, Cande WZ (2006) Alleles of *afd1* dissect REC8 functions during meiotic prophase I. J Cell Sci 119: 3306–3315

14. Golubovskaya IN, Wang CJ, Timofejeva L, Cande WZ (2011) Maize meiotic mutants with improper or non-homologous synapsis due to problems in pairing or synaptonemal complex formation. J Exp Bot 62:1533–1544

15. Koumbaris GL, Bass HW (2003) A new single-locus cytogenetic mapping system for maize (*Zea mays* L.): overcoming FISH detection limits with marker-selected sorghum (*S. propinquum* L.) BAC clones. Plant J 35:647–659

16. Pawlowski WP, Golubovskaya IN, Cande WZ (2003) Altered nuclear distribution of recombination protein RAD51 in maize mutants suggests the involvement of RAD51 in meiotic homology recognition. Plant Cell 15:1807–1816

17. Hamant O, Golubovskaya I, Meeley R, Fiume E, Timofejeva L, Schleiffer A et al (2005) A REC8-dependent plant Shugoshin is required for maintenance of centromeric cohesion during meiosis and has no mitotic functions. Curr Biol 15:948–954

18. Li J, Harper LC, Golubovskaya I, Wang CR, Weber D, Meeley RB et al (2007) Functional analysis of maize RAD51 in meiosis and double-strand break repair. Genetics 176:1469–1482

19. Bass HW, Nagar S, Hanley-Bowdoin L, Robertson D (2000) Chromosome condensation induced by geminivirus infection of mature plant cells. J Cell Sci 113:1149–1160

20. Kikuchi S, Tanaka H, Wako T, Tsujimoto H (2007) Centromere separation and association in the nuclei of an interspecific hybrid between *Torenia fournieri* and *T. baillonii* (Scrophulariaceae) during mitosis and meiosis. Genes Genet Syst 82:369–375

21. Murphy SP, Simmons CR, Bass HW (2010) Structure and expression of the maize (*Zea mays* L.) SUN-domain protein gene family: evidence for the existence of two divergent classes of SUN proteins in plants. BMC Plant Biol 10:269

22. Howe ES, Clemente TE, Bass HW (2012) Maize histone H2B-mCherry, a new fluorescent chromatin marker for somatic and meiotic chromosome research. DNA Cell Biol (in press)

23. Chen H, Swedlow JR, Grote MA, Sedat JW, Agard DA (1995) The collection, processing, and display of digital three-dimensional images of biological specimens. In: Pawley JB (ed) Handbook of biological confocal microscopy. Plenum, New York, pp 197–210

24. Hiraoka Y, Swedlow JR, Paddy MR, Agard DA, Sedat JW (1991) Three-dimensional multiple wavelength fluorescence microscopy for the structural analysis of biological phenomena. Semin Cell Biol 2:153–165

Analyzing Maize Meiotic Chromosomes with Super-Resolution Structured Illumination Microscopy

Chung-Ju Rachel Wang

Abstract

The success of meiosis depends on intricate coordination of a series of unique cellular processes to ensure proper chromosome segregation. Many proteins involved in these cellular events are directly or indirectly associated with chromosomes, especially those required for homologous recombination. These meiotic processes have been explored extensively by conventional light microscopy. However, many features of interest, such as chromatin organization, recombination nodules, or the synaptonemal complex are beyond the resolution of conventional wide-field microscopy. Moreover, in most sample preparation techniques for light microscopy, meiotic cells are squashed, which destroys the spatial organization of the nucleus. Here, I describe a protocol to analyze maize meiotic chromosomes by three-dimensional structured illumination microscopy (3D-SIM), a recently developed high-resolution microscopy technique. This protocol can be used to examine protein localizations at a high resolution level by immunofluorescence.

Keywords *Zea mays*, Meiosis, Chromosome, Super-resolution imaging, Structured illumination microscopy

1 Introduction

Meiosis is a specialized cell division required for all organisms with a sexual cycle. Several unique events occur during meiosis, such as formation of the telomere bouquet, interhomolog recombination initiated by double-strand breaks, pairing and synapsis of homologous chromosomes during prophase I, and monopolar orientation of kinetochores at metaphase I (1–4). These events mediate chromosome behavior and ensure their correct segregation. Many proteins involved in these cellular events have been identified and studied with various approaches (5–7). Their localization, as well as interpretation of phenotypes caused by defects in these proteins in meiotic mutants, relies heavily on cytological analysis, mostly using light microscopy. However, many structures involved in meiosis events are smaller than the resolution limit of conventional

Wojciech P. Pawlowski et al. (eds.), *Plant Meiosis: Methods and Protocols*, Methods in Molecular Biology, vol. 990, DOI 10.1007/978-1-62703-333-6_7, © Springer Science+Business Media New York 2013

light microscopy (8). For example, fine structures of recombination nodules, kinetochores, meiotic chromatin, or the synaptonemal complex cannot be resolved using light microscopes.

The resolution limit of light microscopy can be calculated using the Rayleigh criterion as $0.61\lambda/NA$, where λ represents the wavelength of emitted light and NA is the numerical aperture of the objective lens (9). In practice, the best conventional light microscopes can achieve a lateral (X, Y) resolution of 200–250 nm, whereas axial (Z) resolution is worse. In confocal microscopy, the axial resolution is better than in the traditional wide-field microscopy; however, it is still three times lower than the lateral resolution (10). Two points that are closer to each other than the microscope resolution limit will appear as a single, larger point under the microscope, and no improvements in lens quality or magnification can overcome this problem. Consequently, there has been a long-standing interest in developing new microscopy methods with better resolution.

A number of novel light microscopy systems have been developed in the past few years, including stimulated emission depletion (STED) microscopy (11, 12), stochastic optical reconstruction microscopy/photoactivated localization microscopy (PALM/STORM) (13–17), and three-dimensional structured illumination microscopy (3D-SIM) (18–22). Among them, the PALM/STORM system reaches the best resolution of 20–30 nm in the lateral dimensions and 60 nm in the axial dimension. However, PALM/STORM requires special photoactivatable or photoswitchable fluorescent dyes. 3D-SIM achieves the lateral and axial resolution of 100 nm and 250 nm, respectively (21). It can be used with a wide range of fluorescent dyes and proteins, and the labeling techniques are similar to those used in conventional microscopy. Recently, live imaging of whole cells at a high resolution using structured illumination was also demonstrated (22).

The principle of 3D-SIM super-resolution imaging is based on the Moiré effect (23). 3D-SIM illuminates a sample with a stripe-patterned light generated by a diffraction grating. The structured illumination (i.e., the stripe-patterned light) interacts with the sample's fluorophore, and creates interference fringes, which contain information that can be decoded to reconstruct a sample image with a higher resolution (21). The grating is shifted in five steps and also rotated to three positions at 60-degree intervals. As a result, 15 raw images are captured for each focal plane. After collecting a series of optical sections, a 3D image is reconstructed by computational analysis. For meiotic research, improved resolution of 3D imaging is particularly useful because it provides the means to explore complex structures in 3D-preserved nuclei (24). In this chapter, I present a protocol to study maize meiotic chromosomes with 3D-SIM, including two methods for slide preparation. One method, originally developed in John Sedat (UCSF) and Zacheus

Cande (UC Berkeley) labs, is designed to better preserve 3D chromosome structure (25, 26). The other method, adapted from a protocol used in Rebecca Heald lab (UC Berkeley), is used for experiments in which 3D architecture preservation is not required (27). These methods can be also applied to other plant samples with modifications.

2 Materials

2.1 Sample Fixation

1. Maize anthers dissected from immature male tassels (see Note 1).

2. 10× Buffer A salt solution: 150 mM PIPES, 0.8 M KCl, 0.2 M NaCl, 20 mM EDTA, 5 mM EGTA, adjust pH to 7.0, sterilize by filtration and store at 4°C.

3. 2 M sorbitol: sterilize by filtration and store at 4°C.

4. 0.4 M spermidine in 50 mM PIPES, pH 7.0. Sterilize by filtration. Aliquot and store at −20°C.

5. 0.4 M spermine in 50 mM PIPES, pH 7.0. Sterilize by filtration. Aliquot and store at −20°C.

6. 1 M Dithiothreitol (DTT) in 10 mM sodium acetate, pH 5.2. Sterilize by filtration. Aliquot and store at −20°C.

7. 2× Buffer A. Mix 10 ml of 10× Buffer A salt solution, 16 ml of 2 M sorbitol, 24 ml of sterile water, 50 µl of 0.4 M spermine, 125 µl of 0.4 M spermidine, and 100 µl of 1 M DTT. Store at 4°C for up to 2 weeks.

8. 16% paraformaldehyde (electron microscopy grade).

9. Small petri dish (D × H: 35 mm × 10 mm).

10. Fine forceps (Fine Science Tools, Foster City, CA, USA).

11. Angled probes (Cat. # 10140-03, Fine Science Tools, Foster City, CA, USA).

2.2 Isolation of Meiocyte Suspension

1. Dissecting microscope.

2. Fine forceps (Fine Science Tools, Foster City, CA, USA).

3. Micro knives (Cat. # 10316-14; Fine Science Tools, Foster City, CA, USA).

4. Petri dish (D × H: 90 mm × 10 mm) and parafilm.

5. 1× Buffer A (see Subheading 2.1, item 7).

6. 20 µg/ml solution of bovine serum albumin (BSA).

7. 10× enzyme buffer: 40 mM citric acid, 60 mM sodium citrate (pH 4.8).

8. Enzyme cocktail: 2% Cellulase RS (Yakult Pharmaceutical Industry Co., Ltd., Tokyo, Japan), 1% Macerozyme R-10

(Yakult Pharmaceutical Industry Co., Ltd, Tokyo, Japan), 1% Cytohelicase (Sigma Aldrich, St. Louis, MO, USA) in 1× enzyme buffer. Sterilize by filtration. Aliquot and store at −20°C.

2.3 Slide Preparation Method 1: Acrylamide Sandwich

1. 1 ml of 20% (w/v) ammonium persulfate in water. Prepare immediately before use.

2. 1 ml of 20% (w/v) sodium sulfite, anhydrous. Prepare immediately before use.

3. 15% acrylamide/bisacrylamide (37.5:1) in 1× Buffer A.

4. 2× Buffer A (see Subheading 2.1, item 7).

5. Coverslip (No. 1.5, High performance 18 mm × 18 mm cover glass, Zeiss, Oberkochen, Germany) (see Note 2).

2.4 Slide Preparation Method 2: Cytospin

1. 10× PBS buffer: 80 g of sodium chloride, 2 g of potassium chloride, 14.4 g of disodium phosphate, 2.4 g of potassium diphosphate in 1 L water. pH to 7.4. Sterilize by autoclaving.

2. Cushion solution: 30% glycerol in 1× PBS.

3. Poly-L-lysine coated coverslip (No. 1.5, round coverslip, 12 mm) (see Notes 2 and 3).

4. 15 ml Corex glass tube with self-designed inset fixed in the bottom (see Fig. 1 and Note 4).

5. Custom-designed coverslip adapters (Fig. 1).

6. A chemical spatula with a narrow-flat end.

7. A centrifuge with a swinging-bucket rotor.

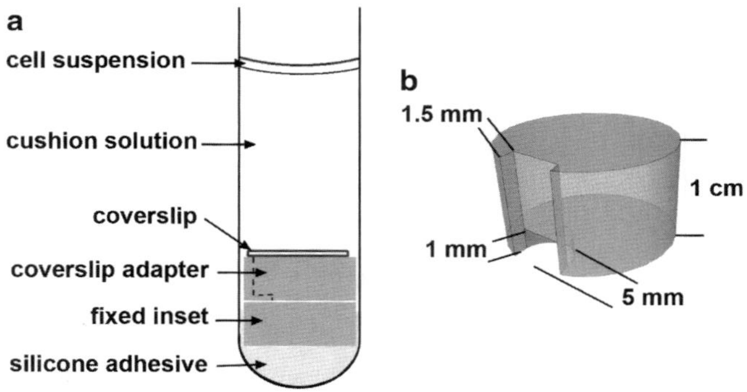

Fig. 1 The cytospin assembly. (**a**) One inset is glued at the bottom of the tube using silicone adhesive. When the cell suspension is ready, add cushion solution, a coverslip adapter, and a coverslip. Then, layer the cell suspension on the *top*. (**b**) To make a coverslip adapter, cut a 1.5 mm indentation on the side of the cylinder and a semicircle indentation (5 × 1 mm) at the *bottom*

2.5 Immunostaining and DAPI Staining

1. 6-well tissue culture plate.

2. 1× PBS (see Subheading 2.4, item 1).

3. Permeabilization buffer: 1× PBS, 1% Triton X-100, 1 mM EDTA.

4. Blocking buffer: 1× PBS, 3% BSA, 0.1% Tween 20, 1 mM EDTA.

5. Wash buffer: 1× PBS, 0.1% Tween 20, 1 mM EDTA.

6. Primary and secondary antibodies (see Note 5).

7. DAPI: 1 mg/ml DAPI (4′,6-diamidine-2-phenylindole dihydrochloride) stock in water.

8. Mounting medium: ProLong Gold antifade (Invitrogen, Carlsbad, CA, USA). Or make your own mounting medium: 5% *n*-propyl gallate, 0.25% 1,4-diazobicyclo-(2,2,2)-octane (DABCO), 0.0025% para-phenylenediamine (PPD), 86% glycerol, and 9% of 0.2 M Tris-HCl buffer (pH 8.0). Wrap the tube completely in foil to protect from light. Store at −20°C.

3 Methods

Critical issues for good quality 3D-SIM images are good sample preservation, a specific, bright, and photostable fluorescent signal, a high signal-to-background ratio, and placing the sample close to the coverslip (see Note 6). The fixation conditions may need to be optimized for sample type and/or primary antibody. The choice of a highly specific primary antibody together with photostable secondary antibody is very important. In this protocol, meiocytes are released from maize anthers and lightly digested with an enzyme cocktail to separate nuclei from the cytoplasm. This step reduces background and increases epitope accessibility since nonspecific binding of the antibody mostly takes place in the cytoplasm. Together, these procedures provide a higher signal-to-background ratio and minimize reconstruction artifacts.

I describe here two protocols for slide preparation, the "acrylamide sandwich" and the "cytospin" methods. The former gives better 3D chromosome preservation, whereas the latter provides better epitope accessibility and helps positioning the sample right next to the coverslip (see Note 6). An example of imaging quality that can be achieved using the "acrylamide sandwich" method and 3D-SIM microscopy is illustrated in Fig. 3.

3.1 Sample Fixation

1. Harvest immature tassels (see Note 1) and fix them as soon as possible after harvesting.

2. Dissect appropriately staged anthers (28) using fine forceps and angled probes. Transfer anthers to a 35 mm petri dish

containing 2 ml of 1× Buffer A. Once enough anthers are collected, add 1 ml of 2× Buffer A and 1 ml of 16% formaldehyde to a final concentration of 4% formaldehyde in 1× Buffer A. Fix anthers for 30 min with gentle shaking.

3. Transfer anthers to a new petri dish containing 4 ml of 1× Buffer A, and wash three times.

4. Use anthers immediately or seal petri dishes with parafilm and store at 4°C (see Note 7).

3.2 Isolation of Meiocyte Suspension

1. Prepare a working surface by placing a piece of parafilm in a 9 cm petri dish. Place 50 µl of 1× Buffer A on the parafilm surface.

2. Under a dissecting microscope, cut off one end of an anther using a micro knife.

3. Carefully transfer the anther into a drop of Buffer A on the parafilm. Hold the anther with forceps at the uncut end and gently press the anther using a needle, from the uncut end towards the cut end to squeeze out meiocytes. A white column of meiocytes will come out from each anther lobe. Discard the remaining anther wall debris.

4. To collect enough meiocytes to make a slide, repeat steps 2 and 3 for five to ten anthers. Use forceps and the micro knife to break meiocyte columns under a dissecting microscope.

5. Coat a 200 µl pipet tip with 20 µg/ml BSA by pipetting it up and down. Use the tip to transfer the meiocyte suspension to a 1.5 ml microfuge tube containing 200 µl of the enzyme cocktail.

6. Mix the meiocytes gently. Let the suspension sit for 15–30 min at room temperature (see Note 8).

7. Centrifuge the meiocytes at $1,000 \times g$ for 3 min to pellet the cells. Discard the supernatant.

8. Add 300 µl of Buffer A and resuspend meiocytes by gentle pipetting.

9. Centrifuge again at $1,000 \times g$ for 3 min and discard the supernatant. Immediately proceed to Subheading 3.3 or 3.4 of slide preparation.

3.3 Slide Preparation Method 1: Acrylamide Sandwich

1. Add 5 µl of Buffer A and resuspend meiocytes with slow pipetting.

2. Prepare 100 µl of 15% acrylamide in 1× Buffer A in a separate microfuge tube for each slide.

3. Work fast until step 8. Transfer 3 µl of meiocyte suspension to the center of a clean coverslip.

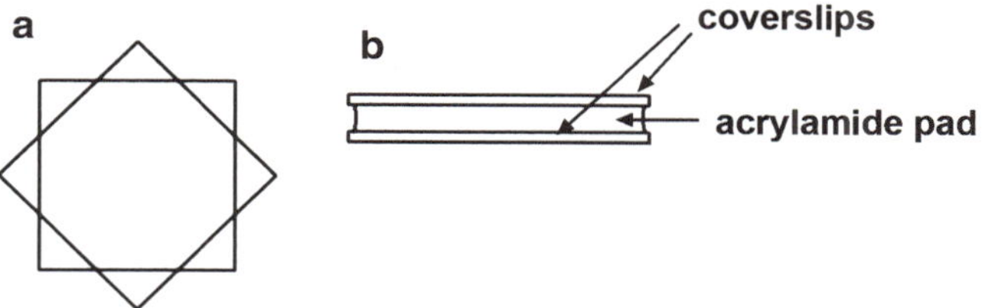

Fig. 2 The acrylamide sandwich. (**a**) Top view. (**b**) Side view

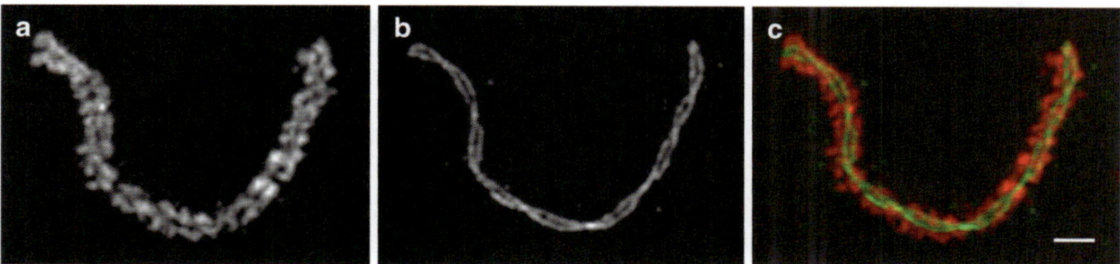

Fig. 3 Maize chromosome prepared with the "acrylamide sandwich method" and imaged by 3D-SIM. (**a**) Chromosome stained with DAPI. (**b**) Visualization of the AFD1 protein, which localizes to the two lateral elements of the synaptonemal complex, which are 140 nm apart. (**c**) Merged image. Scale bar = 1 μm

4. Add 5 μl of 20% ammonium persulfate and 5 μl of 20% sodium sulfite to the tube containing 100 μl of 15% acrylamide in 1× Buffer A. Vortex briefly. Acrylamide will polymerize within 30 s.

5. Immediately add 1.5 μl of the step 4 polyacrylamide mixture into the 3 μl meiocyte droplet on the coverslip.

6. Stir with the pipette tip quickly four to five times. Do not mix by pipetting up and down.

7. Immediately and gently place another coverslip on top of the first coverslip at a 45° angle to make an acrylamide sandwich (Fig. 2). Try to avoid bubbles by lowering the top coverslip carefully and slowly.

8. Let the acrylamide mixture polymerize for 25–30 min.

9. Slowly separate the two coverslips using a razor blade. The acrylamide pad will usually stick to one of the coverslips. Immediately start immunostaining (see Subheading 3.5).

3.4 Slide Preparation Method 2: Cytospin

1. Add 10 μl of 1× PBS to resuspend meiocytes by slow pipetting. Then add another 490 μl of 1× PBS.

2. Prepare a modified COREX tube with a fixed inset (see Note 4). Assemble a cytospin tube as shown in Fig. 1. Add 5 ml of cushion solution, and then drop a coverslip adapter. After the adapter sinks to the bottom of the tube, carefully place a

poly-L-lysine-covered coverslip. The poly-L-lysine side should face up. Push the coverslip gently and let it sink to the bottom of the tube.

3. Gently layer the 500 µl of meiocyte suspension over the cushion solution while slowly rotating the tube.

4. Centrifuge for 20 min at $1,500 \times g$ in a swinging-bucket rotor to deposit meiocytes onto the coverslip.

5. Remove the 80% cushion solution. Carefully insert a small spatula underneath the coverslip adapter through the indentation in the coverslip adapter and slowly lift the adapter and the coverslip out from the tube. Keep track of the side of the coverslip covered with cells. Immediately perform immunostaining (see Subheading 3.5).

3.5 Immunostaining

1. Place coverslips, cells facing up, in 1× PBS, in a 6-well tissue culture plate.

2. Rinse coverslips twice with 1 ml of 1× PBS. Always aspirate the liquid along the walls of the well, not from the surface of the coverslip.

3. To increase membrane permeability, add permeabilization buffer and incubate 1 h for acrylamide pads or 10 min for cytospin coverslips.

4. Rinse with 1× PBS.

5. Incubate 2 h at room temperature in freshly prepared blocking buffer. Serum from the same animal as the secondary antibody may be added to reduce slide background.

6. Incubate with the primary antibody diluted in blocking buffer overnight at 4°C. To prevent evaporation, use a petri dish with two layers of wet Whatman paper. Place a piece of parafilm on top of the filter papers and pipette 50 µl of diluted antibody onto the parafilm. Then, place the coverslip (cells facing down) on top of the antibody droplet. Use fine forceps to manipulate the coverslip. Cover the petri dish and seal with parafilm to prevent cells from drying out.

7. After incubation with the primary antibody, return the coverslip to the 6-well tissue culture plate (cells facing up) and wash three times with wash buffer, 30 min. for each wash for the acrylamide pad coverslips, or 10 min. for each wash for the cytospin coverslips.

8. Incubate with the secondary antibody for 2 h at room temperature. Cover with foil to protect from light. When using an antibody for the first time, always include a control with the secondary antibody alone.

9. Return the coverslip to the tissue culture plate and wash as in step 7. Protect the samples from light.

10. Stain DNA with 10 µg/ml DAPI diluted in 1× PBS for 30 min and then wash briefly in 1× PBS.

11. Drain all liquid from the coverslip by gently tapping the edge of the coverslip on a piece of paper tissue. Place a drop of ProLong Gold antifade (see Note 9) or another antifade mounting medium (see Note 10) onto the coverslip and rotate the coverslip for 5 min to spread the medium over the whole coverslip surface. Remove excess medium by vacuum aspiration at one corner of the coverslip. Add another drop of antifade reagent, turn the coverslip upside down (cells facing down), and place it onto a microscope slide.

12. Press the coverslip to squeeze out as much excess mounting medium as possible. Use vacuum aspiration to remove medium as you press down the coverslip. Take care to not allow the mounting medium spread onto the top of the coverslip.

13. Place the slide on a flat, dry surface and keep for 24 or 48 h at room temperature in the dark, for cytospin or acrylamide pad coverslips, respectively.

14. Carefully clean the surface of the coverslip using 95% ethanol and paper tissue. Wipe the coverslip at least ten times to remove dirt spots or any organic contaminants.

15. The edges of the coverslip can be completely sealed with nail polish and the sample can be stored at –20°C or 4°C. Sealing the slide prevents oxidation and extends the life of the sample.

3.6 Microscopes

The first SIM microscope, called OMX (Optical Microscope eXperimental), was designed and built by John Sedat in collaboration with David Agard and Mats Gustafsson (UCSF). This microscope has two imaging modes: one used for super-resolution 3D-SIM and another one, not discussed here, used for fast live cell imaging using conventional optics. The OMX system was subsequently commercialized by Applied Precision, Inc. (Issaquah, WA, USA), as well as Zeiss and Nikon. Currently, SIM microscopes are available from all three manufacturers (Applied Precision OMX-Blaze, Zeiss ELYRA S.1 and PS.1, and Nikon N-SIM).

4   Notes

1. Under most growing conditions, pollen mother cells in maize enter meiosis about 6–8 weeks after seed germination. At this stage, immature tassels are still inside the stem. Cut the stem vertically just below the top node and check the stage of anther development. In each anther, all meiocytes are developmentally synchronized, and the stage of anther development reflects the position of the flower on the tassel. Therefore, the stage of

anther development can be roughly related to anther length. In general, anthers of 1–3 mm in length contain meiocytes.

2. Use only clean no. 1.5 (0.17 mm thick) coverslips. Exact thickness of individual coverslips should be determined with a high precision microcaliper. Clean coverslips by immersing them in 1 M HCl at 50–60°C for at least 4 h, followed by rinsing with lots of ddH$_2$O. Store clean coverslips in 95% EtOH.

3. Coat one side of a round coverslips with 1 mg/ml poly-lysine in ddH$_2$O by applying enough solution to cover the glass surface for 10 min at room temperature and then washing three times with sterile ddH$_2$O. Allow to air-try.

4. The inset and the coverslip adapter are made of a Plexiglas rod. Cut the rod into small cylinders of 1 cm in height. Glue one of the cylinders at the bottom of a 15 ml COREX glass tube prefilled with silicone adhesive (Fig. 1a). Make sure that the inset is horizontal. The coverslip adapter is made from another small cylinder as shown in Fig. 1b. Ensure that you put the same amount of material in each tube, to get a balanced pair for centrifugation.

5. Use highly specific affinity-purified primary antibodies. There is a wide choice of bright photostable secondary antibodies that can be used to fit the available laser lines. I have successfully used Alexa Fluor- (Invitrogen, Carlsbad, CA, USA) and DyLight- (Jackson ImmunoResearch Inc., West Grove, PA, USA) conjugated secondary antibodies. Fluorescent quantum dots may work as well, but I have not tested them. Fluorescent proteins may also be used, but they necessitate using smaller numbers of optical sections and/or shorter exposure times to reduce photobleaching.

6. The 3D-SIM system can image objects located < 30 μm from the surface of the coverslip. However, the focusing depth is limited by spherical aberration and light scattering properties of the sample. Samples closer to the coverslip give better-reconstructed images.

7. Fixed samples can be stored at 4°C in the dark for 1–2 months. However, freshly fixed material gives better results with less background staining.

8. The length of enzyme digestion may need to be adjusted for each sample used. At the end of the digestion period, meiocyte cell walls should separate from the cytoplasm.

9. To protect from photobleaching and minimize changes of the refractive index in the light path, samples should be mounted in a medium that contains an antifade compound and whose refraction matches the refractive index of the objective lens. ProLong Gold antifade is a commercial mounting medium that solidifies after application, which immobilizes cells for

imaging. However, the refractive index of ProLong medium does not stabilize until the medium cures approximately after 100 h from application.

10. ProLong Gold gives good results when immobilizing objects in the slide is essential to prevent sample dfriting during imaging. If sample drifting is not a problem, use of a home-made glycerol-based mounting medium for better image reconstruction.

Acknowledgments

I thank Zac Cande (UC Berkeley) for supporting this protocol development, as well as John Sedat and Jennifer Feng (UCSF) for access to the SIM microscope. I thank Peter Carlton for help in image processing.

References

1. Harper L, Golubovskaya I, Cande WZ (2004) A bouquet of chromosomes. J Cell Sci 117:4025–4032

2. Hunter N (2006) Meiotic recombination. In: Aguilera A, Rothstein R (eds) Topics in current genetics: molecular genetics of recombination. Springer, Heidelberg, pp 381–442

3. Zickler D, Kleckner N (1999) Meiotic chromosomes: integrating structure and function. Annu Rev Genet 33:603–754

4. Hauf S, Watanabe Y (2004) Kinetochore orientation in mitosis and meiosis. Cell 119:317–327

5. Rieder CL (1999) Methods in cell biology: mitosis and meiosis. Academic, San Diego, CA

6. Hamant O, Ma H, Cande WZ (2006) Genetics of meiotic prophase I in plants. Annu Rev Plant Biol 57:267–302

7. Mercier R, Grelon M (2008) Meiosis in plants: ten years of gene discovery. Cytogenet Genome Res 120:281–290

8. Toomre D, Bewersdorf J (2010) A new wave of cellular imaging. Annu Rev Cell Dev Biol 26:285–314

9. Abbe E (1873) Beitrage zur Theorie des Mikroskops und der mikroskopischen Wahrnehmung. Archiv Mikroskop Anatomie 9:413–468

10. Gu M (1996) Principles of three-dimesional imaging in confocal microscopes. World Scientific, Singapore

11. Klar TA, Jakobs S, Dyba M, Egner A, Hell SW (2000) Fluorescence microscopy with diffraction resolution barrier broken by stimulated emission. Proc Natl Acad Sci U S A 97:8206–8210

12. Willig KI, Kellner RR, Medda R, Hein B, Jakobs S, Hell SW (2006) Nanoscale resolution in GFP-based microscopy. Nat Methods 3:721–723

13. Betzig E, Patterson GH, Sougrat R, Lindwasser OW, Olenych S, Bonifacino JS et al (2006) Imaging intracellular fluorescent proteins at nanometer resolution. Science 313:1642–1645

14. Hess ST, Girirajan TP, Mason MD (2006) Ultra-high resolution imaging by fluorescence photoactivation localization microscopy. Biophys J 91:4258–4272

15. Rust MJ, Bates M, Zhuang X (2006) Sub-diffraction-limit imaging by stochastic optical reconstruction microscopy (STORM). Nat Methods 3:793–795

16. Egner A, Geisler C, von Middendorff C, Bock H, Wenzel D, Medda R et al (2007) Fluorescence nanoscopy in whole cells by asynchronous localization of photoswitching emitters. Biophys J 93:3285–3290

17. Huang B, Wang W, Bates M, Zhuang X (2008) Three-dimensional super-resolution imaging by stochastic optical reconstruction microscopy. Science 319:810–813

18. Gustafsson MG (2000) Surpassing the lateral resolution limit by a factor of two using structured illumination microscopy. J Microsc 198:82–87

19. Schermelleh L, Carlton PM, Haase S, Shao L, Winoto L, Kner P et al (2008) Subdiffraction

multicolor imaging of the nuclear periphery with 3D structured illumination microscopy. Science 320:1332–1336

20. Gustafsson MG, Shao L, Carlton PM, Wang CJ, Golubovskaya IN, Cande WZ et al (2008) Three-dimensional resolution doubling in widefield fluorescence microscopy by structured illumination. Biophys J 94:4957–4970

21. Gustafsson MG (2005) Nonlinear structured-illumination microscopy: wide-field fluorescence imaging with theoretically unlimited resolution. Proc Natl Acad Sci U S A 102: 13081–13086

22. Shao L, Kner P, Rego EH, Gustafsson MG (2011) Super-resolution 3D microscopy of live whole cells using structured illumination. Nat Methods 8:1044–1046

23. Kafri O, Glatt I (1990) The physics of moiré metrology. Wiley, New York, NY

24. Wang CR, Carlton PM, Golubovskaya IN, Cande WZ (2009) Interlock formation and coiling of meiotic chromosome axes during synapsis. Genetics 183:905–915

25. Urata Y, Parmelee SJ, Agard DA, Sedat JW (1995) A three-dimensional structural dissection of Drosophila polytene chromosomes. J Cell Biol 131:279–295

26. Bass HW, Marshall WF, Sedat JW, Agard DA, Cande WZ (1997) Telomeres cluster de novo before the initiation of synapsis: a three-dimensional spatial analysis of telomere positions before and during meiotic prophase. J Cell Biol 137:5–18

27. Wignall SM, Deehan R, Maresca TJ, Heald R (2003) The condensin complex is required for proper spindle assembly and chromosome segregation in *Xenopus* egg extracts. J Cell Biol 161:1041–1051

28. Chang MT, Neuffer MG (1994) Chromosomal behavior during microsporogenesis. In: Freeling M, Walbot V (eds) The maize handbook. Springer, New York, NY, pp 460–475

Chapter 8

Live Imaging of Chromosome Dynamics

Moira J. Sheehan, R. Kelly Dawe, and Wojciech P. Pawlowski

Abstract

Progression of meiosis has been traditionally reconstructed from microscopic images collected from fixed cells. However, studies conducted in a number of species, including plants, indicate that this approach has clear shortcomings in accurately portraying the dynamic nature of meiotic processes. Here, we describe two methods to study chromosome dynamics in live meiocytes in maize, a protocol to observe chromosomes during meiotic prophase I and a technique to monitor chromosome segregation in anaphase I and anaphase II. The first method relies on culturing intact maize anthers and observing meiocytes embedded in the anthers with multiphoton excitation (MPE) microscopy. This approach circumvents difficulties in culturing isolated prophase I meiocytes in plants. The second technique uses culturing isolated meiocytes, which is possible with anaphase cells. Both methods can be fairly easily adapted for use in other plant species. We also detail the kinds of time-lapse movies that can be captured and analyzed using this technique, and describe software that can be utilized for analysis of movies chromosome dynamic in live meiocytes.

Keywords Meiosis, Chromosomes, Chromosome dynamics, Prophase, Anaphase, Cytology, Live imaging, Microscopy

1 Introduction

Reconstructions of the progression of meiotic prophase have been traditionally conducted using fixed meiocytes. Although these reconstructions provided information on the general patterns of chromosome behavior, live imaging studies conducted in a number of species (1–6) indicate that observations of fixed cells are not able to convey the dynamics and complexity of chromosome behavior in live meiocytes. Extensive live microscopy studies of chromosome dynamics have been conducted in unicellular fungi (5, 7). However, meiocytes in higher plants are more difficult targets for live imaging as they are normally embedded in multicellular reproductive structures. In this chapter, we describe two methods to conduct observations of chromosomes in live meiocytes of maize: a protocol to examine chromosome dynamics during early substages of meiotic prophase I using intact cultured anthers (4)

Wojciech P. Pawlowski et al. (eds.), *Plant Meiosis: Methods and Protocols*, Methods in Molecular Biology, vol. 990, DOI 10.1007/978-1-62703-333-6_8, © Springer Science+Business Media New York 2013

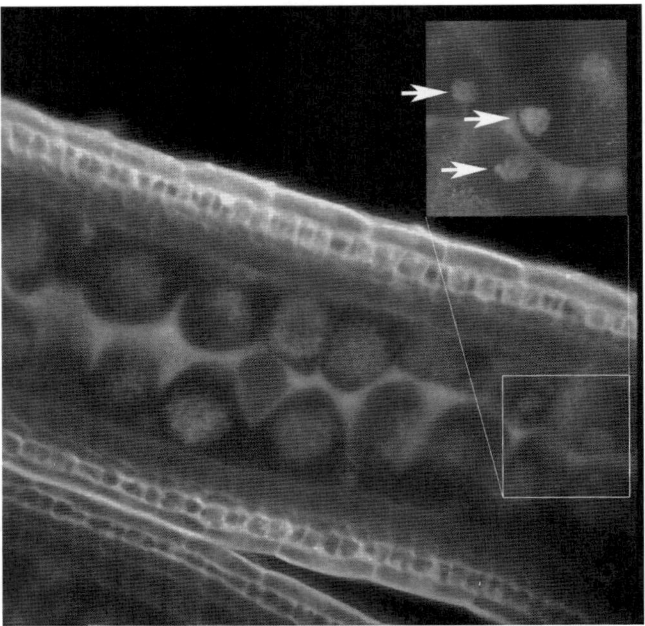

Fig. 1 An optical cross-section through a living maize anther generated with MPE microscopy. The anther was stained with DAPI (*pseudo-colored red*) and Mitotracker Green FM (*green*). A fragment of the anther locule in magnified in *inset. Arrows* point to three meiocyte nuclei at the zygotene stage

and a technique to monitor chromosome behavior during their segregation in anaphase I and II in cultured isolated meiocytes that was described by Yu et al. (1).

In the prophase I live imaging protocol, intact anthers are cultured (8) and the meiocytes are viewed using multiphoton excitation (MPE) microscopy (9), which can penetrate through the surrounding support tissue. The center of the anther locule, where maize meiocytes develop, lies roughly 70–100 μm from the anther surface (Fig. 1). While this depth is beyond the capabilities of confocal microscopy, it is within the demonstrated range of ~200 μm of MPE (9). Using this technique, Sheehan and Pawlowski found that meiotic chromosomes in maize at zygotene and pachytene exhibit extremely vigorous motility (4). In zygotene, the chromosomes exhibited short-range movements, whereas in pachytene, there were slower sweeping motions of large chromosome segments. In addition to the motility of individual chromosome segments, the entire chromatin mass within the nucleus was subject to oscillating rotational motions. This method may also be applicable to later stages of meiosis but it has not yet been tested for those applications.

In the later stages of meiosis, beginning at metaphase I and extending through the remaining stages, meiocytes can be extruded from anthers and observed directly. Separating the cells from their support tissue makes it possible to use simpler microscopy methods since great depth of focus is not required. The method of Yu et al. (1) has been used to measure the rates of anaphase segregation in both meiosis I and II (1, 10). While the isolated meiocyte protocol can be very powerful, cells in prophase I or earlier cannot be extruded without damaging them. Even at later stages when meiocytes can be extruded, few of the isolated cells survive. The methods described here are reliable, but are probably best viewed as starting points, as both protocols can no doubt be improved with further study.

2   Materials

2.1   Plants
(See Note 1)

To ensure uninterrupted availability of anthers, plants should be grown at regular intervals in a growth chamber or greenhouse. The time from planting to meiosis varies by genotype and is affected by environmental conditions. In maize, all anthers on the plant usually enter meiosis within less than a week of each other.

2.2   Reagents and Materials for Initial Anther Staging and Male Inflorescence Harvesting

1. Farmer's fixative—used to fix anthers for initial staging: a 3:1 mixture of 100% ethanol and glacial acetic acid.

2. Acetocarmine—used to stain chromosomes for a quick determination of meiosis stage (see Note 2): dissolve 2% acetocarmine powder in 45% acetic acid and boil 6–8 h in a flask with an attached reflux column. Filter through filter paper when the solution is still warm. Store in a dark bottle.

3. Needle-nosed forceps.

4. Dissecting needles.

5. Razor blades or scalpels.

6. Glass scintillation vials.

7. Glass microscope slides and coverslips.

8. Dissecting stereomicroscope.

9. Wide-field light microscope.

2.3   Reagents and Materials for Live Imaging of Intact Anthers at Prophase I

1. Artificial Pond Water (APW)—anther culture medium: 0.1 M NaCl, 0.1 M $CaCl_2$, and 0.1 M KCl, mixed, filter-sterilized, and stored at room temperature (see Notes 3 and 4).

2. DAPI solution—chromatin stain: make a concentrated stock of 50 mg/ml in water. Add to APW on the day of anther harvest for a final concentration of 50 µg/ml (see Notes 5–7). Store stock at −20°C.

3. DMSO—enhances penetration of stains and other reagents through anther tissues: use at final concentration of 1% for

DAPI staining and no more than 5% for any other application (see Note 8). Store stock at –20°C. Add to APW on the day of anther harvest.

4. Rhodamine 123 (Invitrogen, Carlsbad, CA, USA)—mitochondrial activity stain used to monitor cell viability: make a stock solution of 20 mM in DMSO, store at –20°C. Use at a final concentration of 20 μM. Rhodamine 123 will generate green fluorescence in actively respiring mitochondria (see Note 9).

5. Paraformaldehyde fixative—used for more precise staging of the third anther in each maize floret from which anthers are collected for live imaging (see Note 2): 4% electron microscopy-grade paraformaldehyde in buffer A (see item 6 below), 50 μg/ml DAPI, and 1% DMSO. To fix anthers, incubate them in the fixative for 1 h with gentle shacking. After incubation, replace with buffer A without paraformaldehyde and wash the anthers for 1 h.

6. Buffer A (11):
   (a) 20% of 10× Buffer A salts (150 mM PIPES, 800 mM KCl, 200 mM NaCl, 20 mM EDTA, 5 mM EGTA in water, pH with 1 M NaOH to 6.8, store at 4°C).
   (b) 0.1% spermine stock (0.4 M spermine tetra HCl; Sigma Aldrich, St. Louis, MO, USA) in 50 mM PIPES, store at –20°C.
   (c) 0.25% spermidine stock (0.4 M spermidine; Calbiochem, San Diego, CA, USA) in 50 mM PIPES, store at –20°C.
   (d) 0.2% DTT stock (1.0 M dithiothreotol; Calbiochem, San Diego, CA, USA) in 0.01 M sodium acetate, pH 5.2, store at –20°C.
   (e) 32% sorbitol stock (2 M solution in water, store at 4°C).
   (f) Water.
   (g) Filter-sterilize and store at 4°C.

7. Culture chamber microscope slides, sterile. We use the 8-chamber culture slides (Fig. 2) (Nunc-155411; Thermo Fisher Scientific, Waltham, MA, USA). One to two anthers can be placed in each chamber filled with ~200 μL of culture medium. Chamber slides are also available with a single chamber, 2, 4, and 16 chambers. Prior to anther collection, the chambers should be pre-filled with the desired culture medium.

8. Disposable transfer pipettes, fine-tipped—useful for applying and removing washes and other solutions to and from the culture chamber. Some examples of the pipettes are models 70960-3, -4, and -5 available at http://www.emsdiasum.com/microscopy/products/preparation/pipette.aspx.

9. Adjustable volume micropipettes: used for mixing culture medium components.

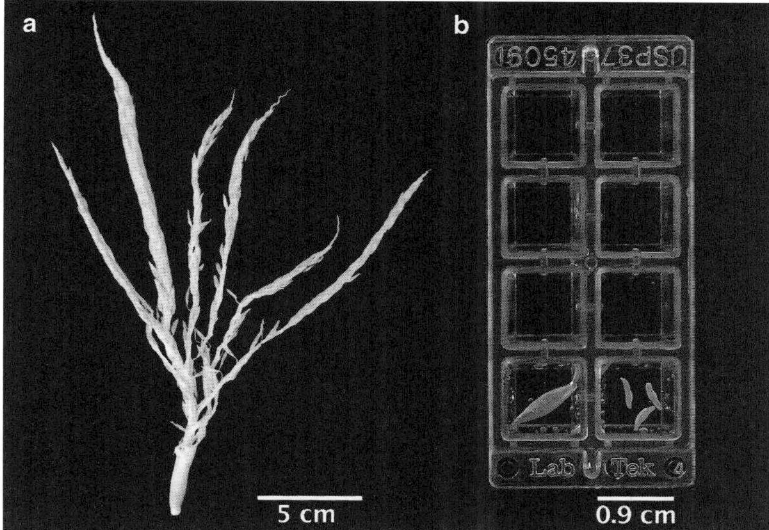

Fig. 2 Culturing maize anthers for live imaging of meiotic prophase I. (**a**) Immature male inflorescence containing anthers with meiocytes at the zygotene stage. (**b**) The 8-chamber culture slide used for live imaging of maize anthers. The upper chamber on the right contains two maize anthers. The lower chamber on the right contains a whole flower

2.4 Reagents and Materials for Live Imaging of Isolated Meiocytes at Anaphase I and II

1. Meiocyte culture medium (see Note 10): 0.3 g/l $Ca(NO_3)_2$, 0.028 g/l $FeSO_3$, 0.08 g/l KNO_3, 0.065 g/l KCl, 0.75 g/l $MgSO_4$, 0.01 mg/l MoO_3, 0.019 g/l NaH_2PO_4, 0.2 g/l $NaSO_4$, 0.050 g/l glutamine, 0.051 g/l glycine, 0.050 g/l L-isoleucine, 0.050 g/l lysine, 0.1 g/l meso-inositol, 0.050 g/l methionine, 0.5 mg/l nicotinic acid, 0.1 g/l pyridoxine, 0.1 mg/l thiamine, 0.050 g/l threonine, 0.050 g/l valine, 0.25 mM n-propyl gallate, 0.1 M sucrose, pH 5.8–5.9. Filter-sterilize and store at –20°C.

2. Cell viability stain: 5 mM Calcein AM (Invitrogen, Carlsbad, CA, USA) in meiocyte culture medium (see Note 11).

3. SYTO 12 fluorescent DNA stain (Invitrogen, Carlsbad, CA, USA) (see Note 7).

4. Culture chamber microscope slides (sterile). Any form of deep-well depression slide will probably work as long as an inverted microscope is used. For the published work, we used single-depression slides (catalog no. 12-560A; Thermo Fisher Scientific, Waltham, MA, USA).

3 Methods

3.1 Initial Anther Staging and Male Inflorescence Harvest (See Note 1)

In maize and other grasses, immature flowers containing cells undergoing meiosis are not visible without dissection, and determination of meiotic stage is required prior to plant harvesting. Determining the developmental stage of male meiocytes in maize is not difficult but requires some practice. To stage and harvest anthers for live imaging of meiosis, we use the following protocol:

1. At the time of meiosis, the immature tassel (the male inflorescence) is still inside the stalk. The presence of the tassel can be felt just below the top node of the plant by gently squeezing the leaf whorl.

2. After establishing that the tassel is large enough to be felt at the first node, make a small incision with a razor blade through the leaves to the tassel, just below the top node.

3. Remove several flowers with needle-nosed forceps into a glass scintillation vial containing Farmer's fixative.

4. Dissect anthers from the collected flowers on a microscope slide under a stereo dissecting microscope.

5. Add a drop of acetocarmine solution for staining (see Note 2). Mix anthers with the stain using a dissecting needle over gentle heat until the color of the stain turns from deep red to purple. Place a coverslip over the anthers and gently press on to break the anthers and release meiocytes. Determine the stage of meiosis under a wide-field light microscope as described by Golubovskaya et al. (12).

6. If the anthers are not yet at the desired meiosis stage, tape over the incision with masking tape and repeat the staging procedure in a day or two.

7. Harvest plants with tassels containing anthers at the desired stage of meiosis at several nodes below the tassel and take to the laboratory for anther dissection.

8. Gently remove leaves surrounding the tassel. To prevent the tassel from drying out, place the tassel on wet paper towels in a tray and put more wet paper towels on top of it.

3.2 Live Imaging of Meiotic Prophase I in Intact Maize Anthers

3.2.1 Preparing Maize Anthers for Live Imaging (See Note 1)

1. After locating florets that contain anthers at the desired stage of meiosis, collect several anthers for the live imaging experiment. Be careful not to puncture or damage the anthers, as it may cause the meiocytes to die. In maize, each floret has three large anthers, which develop synchronously. We always use two of the three anthers for live imaging. Precise meiosis stage determination is more difficult in live anthers than in fixed anthers, and therefore, the third anther from each floret is placed in a paraformaldehyde fixative for more accurate staging (see Note 2).

2. Place the dissected anthers in a microscope chamber slide containing ~200 μl of anther culture medium (Fig. 2) (see Notes 3 and 4).

3. Stain the anthers with a solution of DAPI in the APW medium for 1 h to allow good stain penetration of the inside of the anther (see Notes 5–7).

4. Replace the stain solution with DAPI-free APW (see Notes 9 and 12).

5. The chamber slide should be protected from light whenever possible to minimize photobleaching of fluorophores. Imaging can begin immediately.

3.2.2 Image Capture Using Multiphoton Excitation Microscopy

The exact multiphoton excitation (MPE) settings and acquisition method will vary depending on the microscope workstation used. In our MPE system, we used a TI:Sapphire laser tuned to 780–800 nm to detect both DAPI and Rhodamine 123.

1. Firmly place the micro-chamber slide on the microscope stage with clips. Leave the chamber lid on to minimize dust contamination. Prior to imaging, it may be useful to first locate the anthers inside the culture chamber using the eyepieces rather than the image acquisition software.

2. MPE Images can be collected at one or more optical sections through the 3D object using one or more wavelengths. 2D images taken at a single Z section and at a single time point are useful for measuring objects (lengths, areas, diameters, etc.), and for acquiring wide views of tissue structures (Fig. 1). To capture and analyze chromosome dynamics, we found that the most informative movies contained 20–30 time-lapse images collected at 10-, 20-, and 30-s intervals at a single Z plane. These image numbers and imaging intervals guaranteed no photobleaching or tissue damage, which might occur with more numerous or frequent exposures. Movies covering longer time spans are valuable sources of information on longer-term chromosome dynamics but are often difficult to analyze as cells often move in the microscope field of view after several minutes of imaging. If this happens, post-image analysis may help obtaining useful data from these movies.

3. Test several different combinations of image numbers, image capture intervals, Z section numbers, etc. for every different imaging goal, MPE workstation type, meiosis stage, species, etc.

3.3 Live Imaging of Anaphase I and II in Isolated Meiocytes

3.3.1 Preparing Isolated Meiocytes for Live Imaging

1. Locate florets that contain anthers 1–2 mm in length and place them into a small Petri dish filled with meiocyte culture medium. Pre-staging of anthers by fixation and DAPI staining can be used to determine the sizes of anthers that contain anaphase I and later stages (although with practice it may not be necessary). Very gently cut the end of the anther with a fresh number 15 scalpel, and, as gently as possible, press on the back of the anther with the forceps. Meiocytes will invariably come out from the cut end but they are often damaged and appear as a "string of mush." The youngest meiocytes that can be extruded as separated, undamaged cells are usually in anaphase II.

However, in some genetic backgrounds and on some days, separated and intact meiocytes in anaphase I can be also extruded. Cells should be monitored using Calcein AM since, even in ideal conditions, more than half of the cells will be dead upon extrusion (10).

2. Draw extruded meiocytes into a standard 200 µl pipette tip. Transfer approximately 65 µl of meiocyte suspension into the Fisher 12-560A depression slide. Add SYTO 12 to the medium at a final concentration of 2 mM (see Notes 7 and 8). Place a coverslip over the concavity, leaving a ~2 mm air bubble. The edges of the coverslip can be sealed using rubber cement.

3. Imaging can begin immediately. An FITC filter set can be used to visualize SYTO 12. Since many of the cells will not proceed past the stage observed at extrusion, and because it can take hours for a cell to progress from metaphase to anaphase, it is best to monitor multiple cells at once (but this is only possible with advanced software and a motorized stage). The amount of light a cell is exposed to should be kept to an absolute minimum. In our hands, when total light exposure (from a mercury lamp) for any cell exceeds 1 min, cell viability drops and cells fail to proceed through meiosis.

3.3.2 Image Capture by Standard Epifluorescence Microscopy

Maize chromosomes are large, and isolated meiocytes have little autofluorescence in the FITC channel. A standard fluorescence microscope will probably yield good data, although further processing by deconvolution will sharpen the images. A confocal microscope would further improve image quality, although we have not tested the effects of the laser on isolated cells.

3.4 Analyzing Chromosome Movies Using ImageJ

We use ImageJ (available from: http://rsbweb.nih.gov/ij/), a freeware imaging analysis package, for processing and analyzing live imaging movies.

1. All files are imported to ImageJ as grayscale image stacks. In multiwavelength imaging, each image is a composite of all wavelength channels, which can be pseudo-colored.

2. Prior to analysis, images may be despeckled to improve the signal-to-noise ratio.

3. To improve the ability to identify cellular objects in the images, use the Find Edges algorithm of ImageJ (available under the Process menu). This algorithm analyzes the intensity (brightness) of adjacent pixels to define boundaries of objects, which are placed where the pixel intensity values change. This feature can be used to delineate the nuclear boundaries and identify the position of the nuclear envelope, utilizing the relatively small difference in brightness between the cytoplasm and the nucleoplasm.

4. The ImageStablizer plugin (13) can be used to stabilize images of objects exhibiting jittery movements due to stage vibrations, cellular motions, or for any other reasons. It can also be used to electronically immobilize objects that move laterally across the field of view. Using this plugin, the object is "pinned" down in one place. Doing this allows for analysis of smaller objects moving inside larger objects that also move, for example, fine movements of individual chromosome segments inside nuclei that exhibit rotational movements of the whole chromatin. Relative velocities of the small chromosome segments can be determined without confounding them with the movement of the entire nucleus.

5. For files containing multi-Z image stacks, flat projections can be generated by using a built-in algorithm Z Project and a plugin, Grouped_ZProject (14). Both algorithms generate flat projections from several Z sections in still 3D images or in 3D movies. Flat Z projections are helpful in simplifying 3D images and can be generated based on the average intensity, the maximum intensity, or by a sum of slices.

6. To obtain basic measurements of cellular structures (such as nuclei, nucleoli, or the whole chromatin mass in the nucleus) in still images or in single frames of time-lapse movies, we use the Drawing Tools and the Region-of-Interest (ROI) manager of ImageJ. The drawing tools can also be used to select and straighten curved chromosomes to measure their length.

7. To track and analyze movements of small, single point objects moving in time in time-lapse movies, we use two particle tracking plugins that essentially perform the same task: MTrackJ (15) and ParticleTracker (16). MTrackJ is easier to use and provides the ability to produce publication-quality outputs as it offers more editing options. It also has a more user-friendly interface. The object to be tracked must be manually selected in each time frame. Several objects can be tracked independently in parallel by using differently colored tracks. To do this, one must first track one object, and then do the same for each additional object. The object tracks can be saved as separate movies or can be overlaid over the objects in the originals movies (Fig. 3). Still images with cumulative tracks can also be plugin. Finally, statistics on object velocity and movement directionality through time can be generated.

8. To trace behavior of larger objects, for example, the movements of the entire chromatin mass in the nucleus or shape changes of the nuclear envelope, we utilize the Segmented Line feature. The object is delineated in each time frame and each of the object outlines is saved in the ROI manager. Finally, a time-lapse movie of the outlines is generated, which can be saved separately or overlaid over the originals movie (Fig. 3).

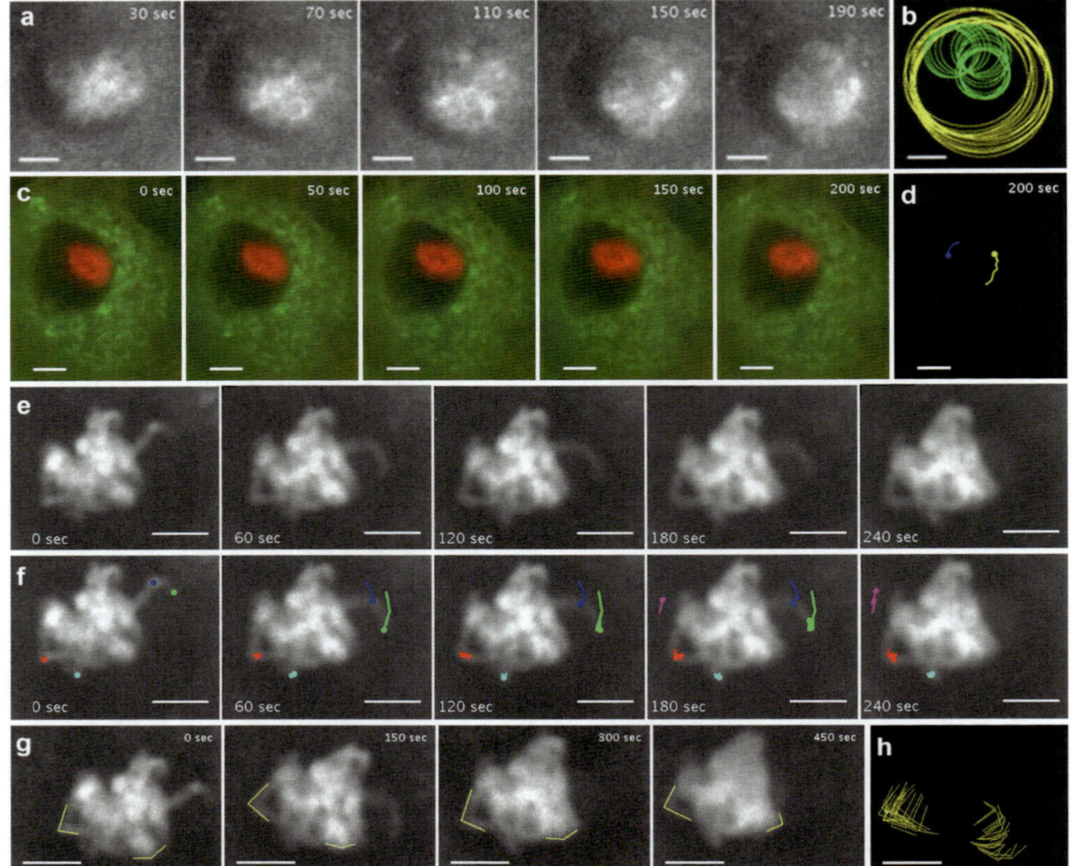

Fig. 3 Patterns of chromosome movements during meiotic prophase I in maize. (**a**) A zygotene nucleus exhibiting rotational motions. (**b**) Cumulative tracks of the nuclear envelope (*yellow*) and nucleolus (*green*) in the nucleus in (**a**) after 190 s. (**c**) A zygotene nucleus showing sliding rotation of the entire chromatin (*red*) along the nuclear envelope. *Green* = cytoplasm stained with Rhodamine 123. (**d**) Cumulative tracks of two anonymous chromosome marks located at the nuclear periphery in the nucleus shown in (**c**) after 200 s. (**e**) Rotational movements of the entire chromatin in a pachytene nucleus. (**f**) The nucleus shown in (**e**) overlaid with trajectories of chromosome marks shown every 60 s for 240 s. *Green* and *blue* mark the chromosome end and an interstitial knob, respectively, of a chromosome arm, which exhibits long-distance sweeping movements. *Red* = a chromosome loop whose both ends are embedded in the chromatin mass. *Cyan* = a stationary chromosome region on the periphery of the chromatin mass. *Magenta* = a free, fast moving chromosome end. (**g**) The nucleus shown in (**e**) with *yellow lines* marking chromatin mass edges. (**h**) Cumulative tracks from (**g**) after 570 s but without chromatin shown. Bars = 5 μm. Adapted from Sheehan and Pawlowski (4)

4   Notes

1. For most meiosis studies in maize, male flowers are used because (a) there are many more male meiocytes than female meiocytes on a plant, (b) the timing of male meiocytes development in maize is synchronized within anthers and florets

and is less variable that that of female meiocytes, and (c) the dissection of male flowers is much easier than female flowers and carries less potential for damage to the meiocytes.

2. The Pawlowski lab uses acetocarmine for quick and rough determination of meiocyte stage, as described in Subheading 3.1, step 5. For precise staging of anthers used for live imaging of prophase I, we fix anthers using paraformaldehyde and conduct 3D microscopy, as described in Chapter 6. The latter method allows very exact determination of meiosis stage, for example, establishing whether meiocytes are at early, mid or late zygotene. Paraformaldehyde fixation is preferred for precise staging as it preserves chromatin and chromosome structure better than the Farmer's fixative. We found live (unfixed) anthers at prophase I difficult to stage directly.

3. Culturing organs for live imaging requires isotonic solutions to maintain tissue viability without altering the speed of growth and development. Furthermore, the medium should be close to optically clear to reduce light scattering. We also tested Murashige and Skoog (MS) medium. However, we found that it had inferior optical clarity compared to APW. Additionally, as APW is a minimal medium (i.e., lacks a carbon source), it is less likely than MS medium to become contaminated by bacteria over the course of experiments.

4. To maintain normal anther growth rate, APW is not supplemented with any growth regulators. Although APW should not be used for long-term culturing (several days or longer), we found that anthers had excellent viability for over 30 h, as evidenced by mitochondrial viability staining (17–19). For example, we observed zygotene meiocytes that progressed to pachytene and pachytene cells that progressed to dyads. Both progression patterns were consistent with the timing of meiosis observed *in planta* (20).

5. Staining live tissues is generally more difficult and often requires higher stain concentrations than staining fixed cells.

6. For MPE microscopy, the fluorophore emission wavelength is an important consideration. Stains with emission in blue or green are the easiest to use in MPE. Stains with yellow and red emissions may also be used but require an MPE system with a laser that can be tuned to an excitation wavelength of 900–950 nm.

7. DAPI was an excellent vital stain for DNA in the anther culture system, but was not effective as a live-cell stain for isolated meiocytes. In the anther culture system, DAPI was more efficient than several SYTO dyes (SYTO 11, SYTO 12, SYTO 13, SYTO 14, SYTO 15, and SYTO 16; Invitrogen, Carlsbad, CA, USA) in penetrating anther walls. For isolated anaphase

meiocytes, SYTO 12 (Invitrogen, Carlsbad, CA, USA) was the most effective chromatin stain. It rapidly penetrates live maize meiocytes to reveal the chromosomes when used at a final concentration of 2 mM. None of the other SYTO stains worked for maize anaphase meiocytes (10), nor did DAPI, Hoechst stains, or propidium iodide (they stained chromosomes but only in dead cells, as assessed using Calcein AM).

8. DMSO is known to polymerize microtubules in vitro. However, it does so when used at concentrations between 8 and 12% (21, 22), which are much higher than the concentrations that we used in live imaging. Consistently, we did not see any difference in either chromosome or cytoplasmic (organellar) motility between DMSO concentrations ranging from 0.1% (which is frequently used in live confocal microscopy experiments (23)) to 5%. It seems likely that DMSO would also facilitate the uptake of DAPI into isolated meiocytes but this has not been tested.

9. In a live imaging system, there is often a need to monitor cell viability in order to distinguish dead cells from cells whose dynamics are disrupted for other reasons. To monitor cell viability, we visualize mitochondrial activity because many live mitochondrial stains are available and can be tried. We tested three dyes that can only fluoresce when within actively respiring mitochondria. Rhodamine 123 provided the most consistent mitochondrial staining of meiocytes of the three stains. We also tried $DiOC_7(3)$ at a final concentration of 200 μM and Mitotracker Green FM, which was tested at a final concentration of 200 nM. $DiOC_7(3)$ stained well the anther epidermal layers but was not visible in meiocytes, suggesting that the stain could not sufficiently penetrate to the inside of the anther. Mitotracker Green FM provided sufficient mitochondrial staining in leptotene meiocytes but staining was not reliably found in zygotene or pachytene meiocytes.

10. Isolated maize meiocytes survive for at least 8 h in a rye meiocyte culture medium described by Pena (24). We use this medium without micronutrients and with a reduced sucrose concentration. In addition, we add 0.25 mM n-propyl gallate, which was shown to increase the longevity of maize protoplasts in culture (25) and showed a measurable increase in viability when cells were exposed to light (10).

11. Calcein AM is converted into a fluorescent product in living cells but does not stain dead cells.

12. The anther culturing system can be easily used to test the effects of various drugs on meiocyte development and chromosome dynamics. Drugs are added to the DAPI-containing anther culture medium for incubation simultaneous with

chromosome staining. After staining, the culture medium should be replaced with medium free of DAPI but with the same concentration of the drug. We used this approach to test the effects of cytoskeleton-disrupting drugs:

(a) Latrunculin B (Lat B): disrupts actin filament polymerization. 100 μM latrunculin B (Sigma Aldrich, St. Louis, MO, USA) stock solution should be stored at −20°C. We used final concentrations of 500 nM and 1 μM.

(b) Colchicine: disrupts microtubule polymerization. 100 mM colchicine (Sigma Aldrich, St. Louis, MO, USA) stock solution should be stored at −20°C. We used final concentrations of 1 mM and 5 mM. For the 1 mM colchicine concentration, we used DMSO at a final concentration of 1% but for the 5 mM colchicine treatment, we used 5% as the final DMSO concentration.

Acknowledgments

Development of the live imaging protocol in the Pawlowski lab was supported by a USDA-NRI postdoctoral fellowship to M.J.S. and a USDA-AFRI grant to W.P.P.

References

1. Yu H-G, Muszynski MG, Dawe RK (1999) The maize homologue of the cell cycle checkpoint protein MAD2 reveals kinetochore substructure and contrasting mitotic and meiotic localization patterns. J Cell Biol 145:425–435

2. Conrad MN, Lee CY, Chao G, Shinohara M, Kosaka H, Shinohara A et al (2008) Rapid telomere movement in meiotic prophase is promoted by *NDJ1*, *MPS3*, and *CSM4* and is modulated by recombination. Cell 133:1175–1187

3. Koszul R, Kim KP, Prentiss M, Kleckner N, Kameoka S (2008) Meiotic chromosomes move by linkage to dynamic actin cables with transduction of force through the nuclear envelope. Cell 133:1188–1201

4. Sheehan MJ, Pawlowski WP (2009) Live imaging of rapid chromosome movements in meiotic prophase I in maize. Proc Natl Acad Sci U S A 106:20989–20994

5. Chikashige Y, Ding D-Q, Funabiki H, Haraguchi T, Mashiko S, Yanagida M et al (1994) Telomere-led premeiotic chromosome movement in fission yeast. Science 264:270–273

6. Baudrimont A, Penkner A, Woglar A, Machacek T, Wegrostek C, Gloggnitzer J et al (2010) Leptotene/zygotene chromosome movement via the SUN/KASH protein bridge in *Caenorhabditis elegans*. PLoS Genet 6:e1001219

7. Scherthan H, Wang H, Adelfalk C, White EJ, Cowan C, Cande WZ et al (2007) Chromosome mobility during meiotic prophase in *Saccharomyces cerevisiae*. Proc Natl Acad Sci U S A 104:16934–16939

8. Cowan CR, Cande WZ (2002) Meiotic telomere clustering is inhibited by colchicine but does not require cytoplasmic microtubules. J Cell Sci 115:3747–3756

9. Denk W, Strickler JH, Webb WW (1990) Two-photon laser scanning fluorescence microscopy. Science 248:73–76

10. Yu H-G, Hiatt EN, Chan A, Sweeney M, Dawe RK (1997) Neocentromere-mediated chromosome movement in maize. J Cell Biol 139:831–840

11. Bass HW, Marshall WF, Sedat JW, Agard DA, Cande WZ (1997) Telomeres cluster de novo before the initiation of synapsis: a three-dimensional spatial analysis of telomere positions before and during meiotic prophase. J Cell Biol 137:5–18

12. Golubovskaya IN, Harper LC, Pawlowski WP, Schichnes D, Cande WZ (2002) The *pam1* gene is required for meiotic bouquet formation and efficient homologous synapsis in maize (*Zea mays*, L.). Genetics 162:1979–1993

13. Li K (2008) The image stabilizer plugin for ImageJ (http://www.cs.cmu.edu/~kangli/code/Image_Stabilizer.html)

14. Holly C (2004) Grouped ZProjector (http://rsbweb.nih.gov/ij/plugins/group.html). Holly Mountain Software

15. Meijering E, Dzyubachyk O, Smal I (2012) Methods for cell and particle tracking. Methods Enzymol 504:183–200

16. Sbalzarini IF, Koumoutsakos P (2005) Feature point tracking and trajectory analysis for video imaging in cell biology. J Struct Biol 151:182–195

17. Johnson LV, Walsh ML, Chen LB (1980) Localization of mitochondria in living cells with Rhodamine 123. Proc Natl Acad Sci U S A 77:990–994

18. Keij JF, Bell-Prince C, Steinkamp JA (2000) Staining of mitochondrial membranes with 10-nonyl acridine orange, MitoFluor Green, and MitoTracker Green is affected by mitochondrial membrane potential altering drugs. Cytometry 39:203–210

19. Emaus RK, Grunwald R, Lemasters JJ (1986) Rhodamine 123 as a probe of transmembrane potential in isolated rat-liver mitochondria: spectral and metabolic properties. Biochim Biophys Acta 850:436–448

20. Hsu SY, Huang YC, Peterson PA (1988) Development pattern of microspores in *Zea mays* L.—the maturation of upper and lower florets of spikelets among an assortment of genotypes. Maydica 33:77–98

21. Pale ek J, Hašek J (1984) Visualization of dimethyl sulphoxide-stabilized tubulin containing structures by fluorescence staining with monoclonal anti-tubulin antibodies. Histochem J 16:354–356

22. Xu C-H, Huang S-J, Yuan M (2005) Dimethyl sulfoxide is feasible for plant tubulin assembly *in vitro*: a comprehensive analysis. J Integr Plant Biol 47:457–466

23. DeBolt S, Gutierrez R, Ehrhardt DW, Melo CV, Ross L, Cutler SR et al (2007) Morlin, an inhibitor of cortical microtubule dynamics and cellulose synthase movement. Proc Natl Acad Sci U S A 104:5854–5859

24. De La Peña A (1986) "*In vitro*" culture of isolated meiocytes of rye, *Secale cereale* L. Environ Exp Bot 26:17–21

25. Cutler AJ, Saleem M, Coffey MA, Loewen MK (1989) Role of oxidative stress in cereal protoplast recalcitrance. Plant Cell Tissue Organ Cult 18:113–128

Chapter 9

Immunolocalization of Meiotic Proteins in *Brassicaceae*: Method 1

Liudmila A. Chelysheva, Laurie Grandont, and Mathilde Grelon

Abstract

Plant meiosis studies have enjoyed a fantastic boom in recent years with the use of *Arabidopsis thaliana* as an important model species for developmental studies because of its small genome, short life cycle, and large mutant collections. Unlike other eukaryotic models, plant meiosis does not display strict checkpoints and rarely commits to apoptotic processes, which makes it possible to investigate the whole meiotic process (spanning from premeiotic interphase to spore formation) in knockout mutants. In this chapter we describe a protocol for immunolabelling *Arabidopsis* and *Brassica* meiotic proteins on robustly spread chromosomes. This protocol allows the detection of a large range of proteins on well-preserved chromosomes and throughout the entire meiotic process.

Keywords *Arabidopsis*, *Brassica napus*, Cytogenetics, Chromosome, Meiosis, Immunolabelling

1 Introduction

Significant progress in the study of *Arabidopsis* meiosis has been made these last years. With the development of cytogenetic techniques (1–3) chromosome organization and behavior could be analyzed during meiotic progression in wild type and various meiotic mutants. These first protocols were based on an acetic acid cell-spreading of flower buds fixed in Carnoy's fixative. Then, Armstrong and colleagues (4) developed techniques for immunolabelling meiotic proteins, which allowed the detailed analysis of spatial and temporal expression of different proteins throughout prophase of meiosis I. This commonly used protocol requires fresh material on which lipsol or Triton X-100 spreading and paraformaldehyde fixation are applied (see Chapter 10). Unfortunately this method of immunolabelling has limitations: often poor resolution because of mild spreading efficiency of lipsol, loss of stages beyond pachytene, and poorly preserved chromosome organization. To overcome these limitations, we developed a technique that allows

Wojciech P. Pawlowski et al. (eds.), *Plant Meiosis: Methods and Protocols*, Methods in Molecular Biology, vol. 990, DOI 10.1007/978-1-62703-333-6_9, © Springer Science+Business Media New York 2013

immunolocalization of proteins using Carnoy's fixative and acetic acid chromosome spreads (5). This method combines a strong fixation to preserve chromosome structure, acetic acid spreading to remove the cytoplasm, and microwave treatment to increase the accessibility of protein to antibodies.

2 Materials

2.1 Plants

1. *Arabidopsis thaliana* plants grown in greenhouse, photoperiod 16 h day and 8 h night; temperature 20°C day and night; humidity 70% (see Note 1).

2. *Brassica napus* plants grown in greenhouse, photoperiod 16 h per day and 8 h per night; temperature 22°C day and 18°C night; humidity 65% (see Note 1).

2.2 Chromosome Preparation

1. Carnoy's fixative: mix absolute ethanol (3 volumes) and glacial acetic acid (1 volume).

2. Citrate buffer pH 4.5: 10 mM sodium citrate/citric acid, pH 4.5. Weigh 2.94 g of tri-sodium citrate and transfer to a graduated glass beaker. Add distilled water up to 800 ml. Prepare 100 ml of 100 mM citric acid (2.1 g in 100 ml of distilled water). Progressively add citric acid solution to the 800 ml of tri-sodium citrate solution until pH 4.5. Top up to 1 l with distilled water. Sterilize and make 40 ml aliquots. Store at –20°C.

3. Citrate buffer pH 6.0: 10 mM sodium citrate/citric acid, pH 6.0. Prepare 0.1 M citric acid (21.01 g in 1 l of distilled water). Prepare 0.1 M tri-sodium citrate (29.41 g in 1 l of distilled water). Add 9 ml of 0.1 M citric acid solution to 41 ml of 0.1 M tri-sodium citrate solution. Top up to 500 ml with distilled water. Store at 4°C (see Note 2).

4. Digestion enzyme mixture: 0.3% (w/v) cellulase Onozuka R-10 (Yakult Pharmaceutical Industry Co., Ltd., Tokyo, Japan), 0.3% (w/v) pectolyase Y-23 (MP Biomedicals, Santa Ana, CA, USA), 0.3% (w/v) cytohelicase (Sigma Aldrich, St. Louis, MO, USA) in 10 mM citrate buffer pH 4.5. Weigh 1.5 g of each enzyme and dissolve in 500 ml of sterilized 10 mM citrate buffer pH 4.5. Aliquot by 40 ml and store at –20°C.

5. 60% acetic acid.

6. Embryo dishes.

7. Stereomicroscope equipped with eyepiece micrometer disk.

8. Standard microscope slides (Knittel Glaser, Braunschweig, Germany) cleaned with 70% ethanol.

9. Dissection needles, a hook (curved dissection needle), and a fine forceps.

10. Moist chamber for enzyme digestion. A hermetic plastic box $180 \times 120 \times 75$ mm^3 with a wet paper on bottom.

11. Glass Pasteur pipette.

12. Heating block (45°C).

13. Incubator (37°C).

14. Glass and plastic Hellendahl jars (VWR, Radnor, PA, USA).

15. Microwave oven.

2.3 Immunolabelling

1. 1× PBS: 10 mM sodium phosphate, pH 7.0, 143 mM NaCl.

2. PBST: 0.1% Triton-X 100 in 1× PBS.

3. PBST-BSA: 1% BSA in 1× PBST.

4. Secondary antibodies conjugated to Alexa (Invitrogen, Carlsbad, CA, USA) or DyLight (Thermo Fisher Scientific, Waltham, Massachusetts, USA).

5. DAPI solution: 2 μg/mL DAPI (4′,6-diamidino-2-phenylindole) in Vectashield (Vector Labs, Burlingame, CA, USA).

6. Laboratory film Parafilm "M" (Sigma Aldrich, St. Louis, MO, USA).

7. Coverslips 24 × 32 mm.

8. Moist chamber: a hermetic container (e.g., square Petri dish) with wet paper on the bottom and two rails in order to keep microscopic slides horizontally.

9. Incubator (37°C).

10. Fluorescence microscope equipped with a high resolution CCD camera.

3   Methods

3.1 Chromosome Preparation

1. Fix the inflorescences in freshly prepared Carnoy's fixative for 3×60 min and keep them in fixative for at least 3 days at 4°C (see Note 3). Fixed material can be stored in fixative at 4°C or −20°C for several months.

2. Rinse the fixed inflorescences once with distilled water in an embryo dish.

3. Replace the water with citrate buffer pH 4.5 and wash 2×5 min.

4. Replace the citrate buffer with the digestion enzyme mixture. Incubate for 3 h (3.5 h for *Brassica* inflorescences) in a moist chamber at 37°C. Ensure that all material is submerged in the digestion mixture.

5. Replace the digestion mix with water and from now on keep material on ice. Digested inflorescences can be stored at 4°C overnight.

6. Take a single inflorescence in a drop of distilled water on a slide and select three flower buds of appropriate stage (see Notes 4 and 5). Extract anthers and place them in 4 μl of water on an ethanol-cleaned slide. For *Brassica* samples, prepare each slide with a single anther.

7. Crush anthers with a hook until a fine suspension has formed (see Note 6).

8. Add 10 μl of 60% acetic acid to the suspension on the slide. Place the slide on the heating block for 2 min at 45°C. Add another 10 μl of 60% acetic acid after the first minute of heating. For *Brassica*, add 20 μl of 60% acetic acid on the slide and place it for 5 min at 45°C. Add 10 μl of 60% acetic acid after 2 min of heating.

9. During step 8, stir the drop with the hook without stopping (see Note 7) in order to remove as much cytoplasm as possible.

10. Fix the cells on slides first by pipetting cold Carnoy's fixative around the drop of the cleared cell suspension. Then, at the end, add the remaining Carnoy's by pipetting directly onto the top of the suspension. Hold the slide vertically to get rid of the liquid. Air-dry the preparation.

11. Put up to eight slides in 10 mM citrate buffer pH 6 in a plastic Hellendahl jar and microwave them for 45 s at 850 W.

12. Transfer the slides immediately to PBS-T in a glass Hellendahl jar. Incubate for 5 min at RT.

3.2 Immunolabelling

1. Dilute the primary antibodies to the appropriate concentration in PBST-BSA and centrifuge them 3 min at $10,000 \times g$.

2. Use 50 μl of the antibody working solution per slide, cover with 24×40 mm pieces of parafilm.

3. Incubate at 4°C for 24–48 h in a moist chamber.

4. Rinse the slides in PBST, 3×15 min at RT.

5. Mix the secondary antibodies in PBST-BSA at a dilution of 1:100 to 1:2,000 (see Note 8).

6. Use 50 μl of solution per slide, cover with 24×40 mm pieces of parafilm and incubate at 37°C for 1 h in a dark moist chamber.

7. Rinse the slides in PBST, 3×10 min at RT.

8. Put 15 μl of DAPI solution per slide and cover with a 24×32 coverslip.

9. Observe the slides under a fluorescent microscope using the appropriate filters (see Note 9) (Figs. 1, 2, and 3).

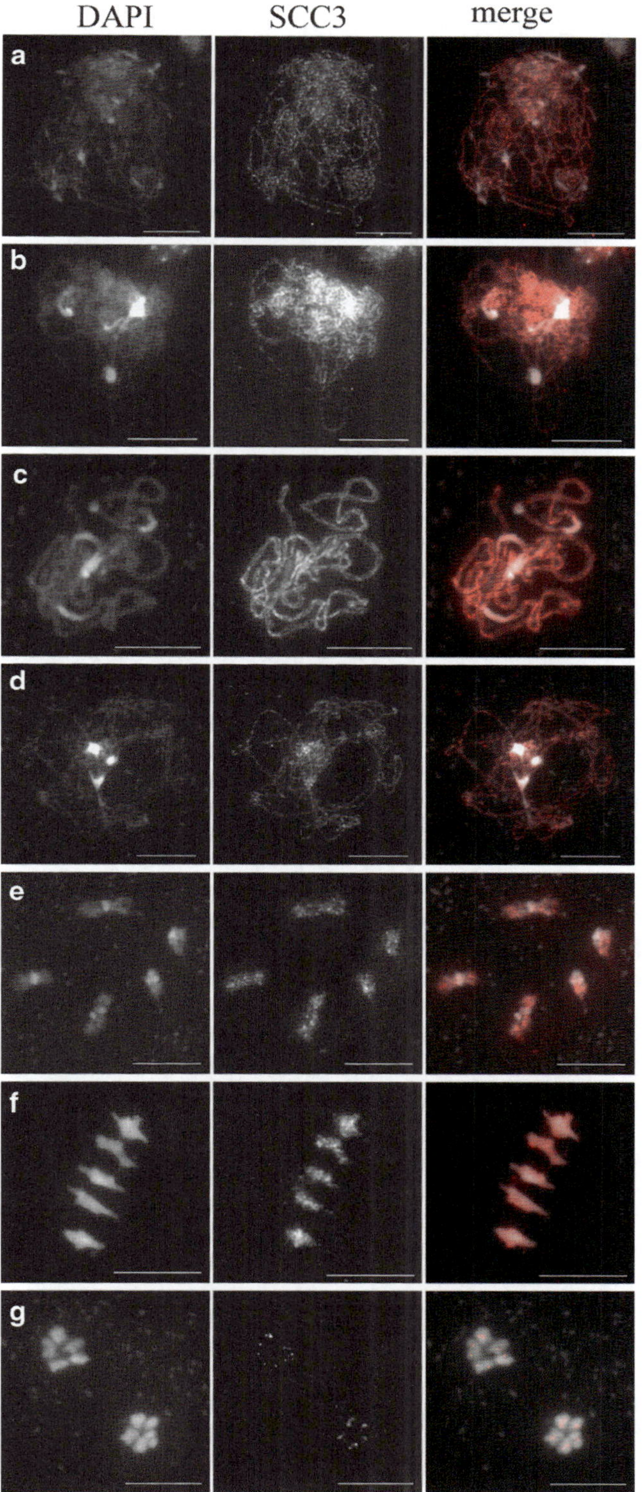

Fig. 1 Immunolocalization of AtSCC3 on *Arabidopsis* pollen mother cells. The cohesin AtSCC3 (6) was immunolocalized on *Arabidopsis* pollen mother cells chromosomes after acetic acid spreading. For each cell DAPI, AtSCC3 and merged signals are shown. (**a**) Leptotene. (**b**) Zygotene. (**c**) Pachytene. (**d**) Diplotene. (**e**) Diakinesis. (**f**) Metaphase I. (**g**) Telophase 1. Bar: 10 μm

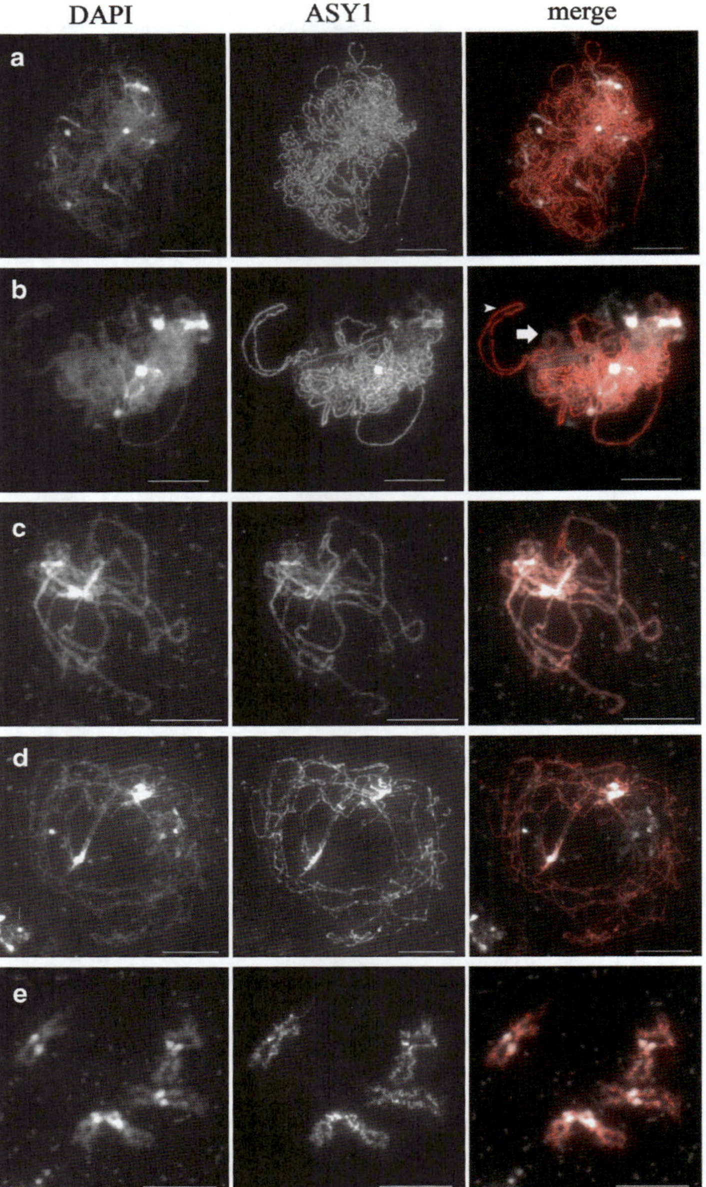

Fig. 2 Immunolocalization of ASY1 on *Arabidopsis* pollen mother cells. The axis-associated protein ASY1 was immunolocalized on *Arabidopsis* pollen mother cells chromosomes after acetic acid spreading (4). For each cell DAPI, ASY1 and merged signals are shown. (**a**) A leptotene cell showing numerous dotted lines corresponding to the ASY1 signal. (**b**) A zygotene cell showing continuous ASY1 signals along chromosomal axes; the *arrowhead* indicates a bright ASY1-labelled unsynapsed region, while the arrow indicates a synapsed region, with faint ASY1 signal. (**c**) Pachytene. (**d**) Diplotene. (**e**) Diakinesis. Bar, 10 μm

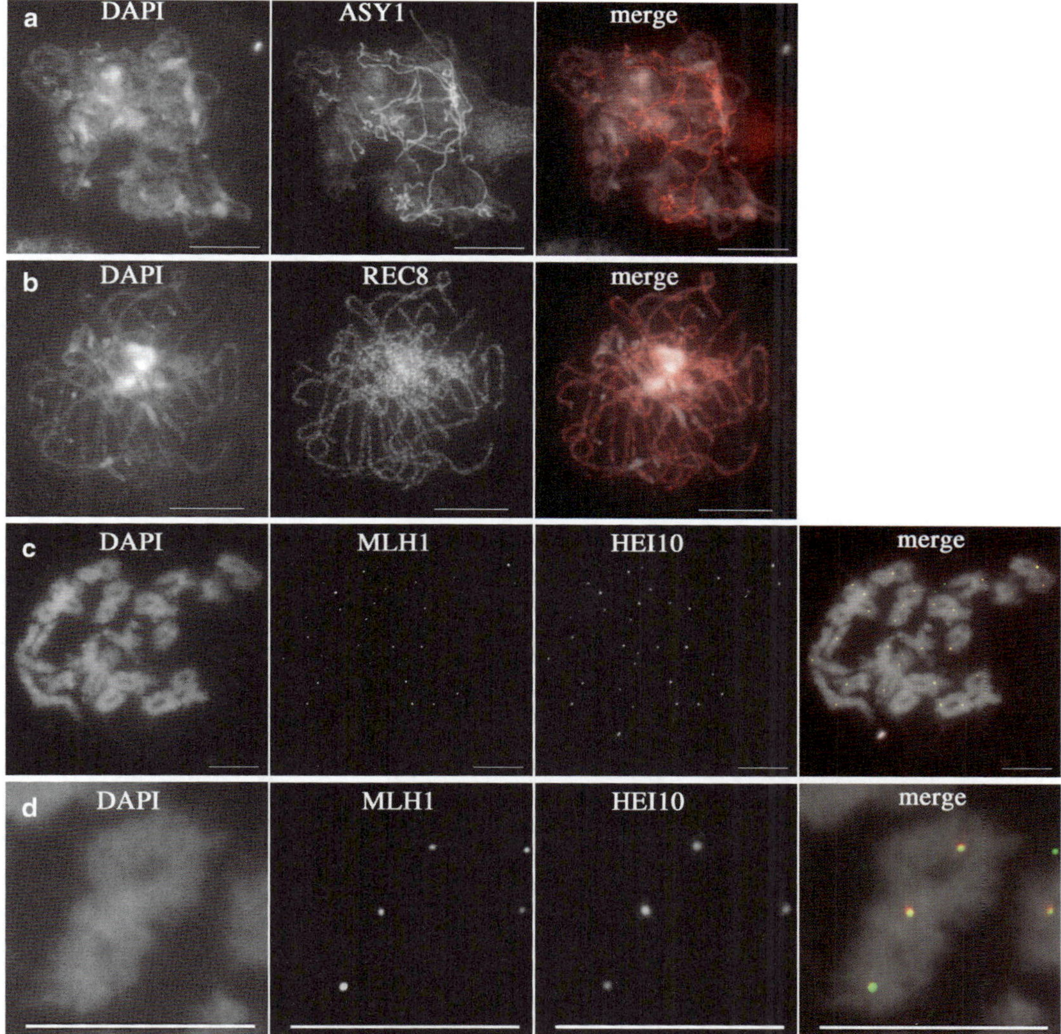

Fig. 3 Immunolocalization of AtASY1, AtREC8, AtMLH1, and AtHEI10 on *Brassica napus* pollen mother cells. (**a**) Immunolocalization of the axis protein ASY1 on *Brassica napus* pollen mother cells chromosomes at zygotene using an antibody directed against the *Arabidopsis thaliana* ASY1 protein (4). DAPI, ASY1 and merged signals are shown. (**b**) Immunolocalization of the meiotic cohesin REC8 on *Brassica napus* pollen mother cells chromosomes at pachytene using an antibody directed against the *Arabidopsis thaliana* REC8 protein (7). DAPI, REC8 and merged signals are shown. (**c**) Co-immunolocalization of MLH1 and HEI10 proteins on *Brassica napus* pollen mother cells chromosomes at diakinesis. Anti-MLH1 and anti-HEI10 sera were developed against the *Arabidopsis thaliana* proteins (5). (**d**) A zoom on one bivalent from (**c**). For (**c**) and (**d**), DAPI, MLH1, HEI10, and merged signals are shown. Bar, 10 μm

4 Notes

1. Good and reproducible growth conditions are particularly important because inappropriate growth conditions can considerably affect the dynamics of meiotic progression. Furthermore, it can be difficult to recover enough material from unhealthy plants.

2. We found that the utilization of commercial citrate buffer pH 6.0 (Diapath S.p.A., Martinengo, Italy) gives very good results.

3. A fixation shorter then 3 days can be insufficient to make all proteins available to immunodetection. In consequence, meiocytes extracted from partially fixed buds will show only partial chromosome immunostaining.

4. In *Arabidopsis*, pollen mother cells develop synchronously, which make it possible to relate the meiotic progression to the bud size. Nevertheless, this correspondence is dependent on the genotype (variability among the ecotypes or between wild type and mutants) and growth conditions. Therefore, it is very important to define it for your own material. Roughly, 0.4 mm buds contain prophase I pollen mother cells, 0.5 mm buds contain metaphase I meiocytes, 0.6 mm buds—metaphase II to telophase II, and 0.75 mm buds tetrads and microspores in Columbia-0.

5. In *Brassica*, bud size is poorly correlated with pollen mother cell development stages but development stages are homogenous among the various anthers of a single bud. Therefore, it is possible to select appropriate buds after acetocarmine staining of one anther per bud. To do it, extract one anther from a fixed bud and place it on a slide. Add 30 µl of acetocarmine and crush the anther. Cover with a coverslip 24 × 32 mm. Place the slide on the heating block for 8 min at 45°C. Observe the slides under a bright field microscope.

6. It is very important to make a fine cell suspension in order to treat all cells efficiently with acetic acid and to get rid of most of the cytoplasm. However, it is also very important to prevent drop evaporation that would make meiocytes stick to the slide and dry, rendering them unavailable to the subsequent acetic acid treatment. In this case, many faintly stained meiocytes will be observed, surrounded by cytoplasm.

7. To amplify acetic acid treatment, the droplet with cells should be stirred vigorously with a hook without touching the glass surface.

8. The appropriate dilution for secondary antibodies should be determined experimentally and can sometimes be more diluted than suggested by the supplier.

9. The fluorescence signal of far-red fluorochromes (e.g., Alexa-647, DyLight-640) can be visualized only by image acquisition.

Acknowledgments

We thank Wayne Crismani for constructive reading of the manuscript, and C. Franklin and C. Makaroff for providing ASY1 and REC8 antibodies, respectively.

References

1. Fransz P, Armstrong S, Alonso-Blanco C, Fischer TC, Torres-Ruiz RA, Jones GH (1998) Cytogenetics for the model system Arabidopsis thaliana. Plant J 13:867–876

2. Heslop-Harrison JS (1998) Cytogenetic analysis of Arabidopsis. Methods Mol Biol 82:119–127

3. Ross KJ, Fransz P, Jones GH (1996) A light microscopic atlas of meiosis in Arabidopsis thaliana. Chromosome Res 4:507–516

4. Armstrong SJ, Caryl AP, Jones GH, Franklin FCH (2002) Asy1, a protein required for meiotic chromosome synapsis, localizes to axis-associated chromatin in *Arabidopsis* and *Brassica*. J Cell Sci 115:3645–3655

5. Chelysheva L, Grandont L, Vrielynck N, le Guin S, Mercier R, Grelon M (2010) An easy protocol for studying chromatin and recombination protein dynamics during Arabidopsis thaliana meiosis: immunodetection of cohesins, histones and MLH1. Cytogenet Genome Res 129:143–153

6. Chelysheva L, Diallo S, Vezon D, Gendrot G, Vrielynck N, Belcram K et al (2005) AtREC8 and AtSCC3 are essential to the monopolar orientation of the kinetochores during meiosis. J Cell Sci 118:4621–4632

7. Cai X, Dong F, Edelmann RE, Makaroff CA (2003) The Arabidopsis SYN1 cohesin protein is required for sister chromatid arm cohesion and homologous chromosome pairing. J Cell Sci 116:2999–3007

Chapter 10

Immunolocalization of Meiotic Proteins in *Arabidopsis thaliana*: Method 2

Susan Armstrong and Kim Osman

Abstract

Advances in the molecular biology and genetics of *Arabidopsis thaliana* have led to it becoming an important model for the analysis of meiosis in plants. Cytogenetic investigations are pivotal to meiotic studies and a number of technological improvements for *Arabidopsis* cytology have provided a range of tools to investigate chromosome behavior during meiosis (Jones et al. Chromosome Res 11:205–215, 2003). This chapter contains a detailed description of an immunological technique currently used in our lab for the preparation of meiotic chromosomes for immunolocalization.

Keywords Immunological techniques, *Arabidopsis thaliana*, Meiosis, Chromosomes

1 Introduction

Immunolocalization of meiotic and chromosome associated proteins to *Arabidopsis thaliana* meiocytes has been carried out using either spreading or squash procedures. Both of these techniques have a number of limitations. For example, with the spreading technique we lose the overall 3D organization of the meiocyte but achieve good, clean immunolocalization with the antibodies in question. On the other hand, although the squash technique retains the organization of the cell, interpretation is often hindered by high background staining due to the surrounding cellular contents. This chapter focuses on a spreading technique that we have developed for *A. thaliana* meiocytes (1–4) and is based on an earlier electron microscope (EM) spreading technique for plant meiocytes, also developed in the Birmingham laboratory (5).

Wojciech P. Pawlowski et al. (eds.), *Plant Meiosis: Methods and Protocols*, Methods in Molecular Biology, vol. 990, DOI 10.1007/978-1-62703-333-6_10, © Springer Science+Business Media New York 2013

2 Materials

2.1 Plants

Sow seeds of *Arabidopsis* in 6 cm diameter pots in soil-based compost. The plants are grown in dedicated growth chambers maintained at 18°C with a 16 h light cycle. In our conditions, the plants arrive at flowering stages at 5–6 weeks.

2.2 Immunocytology: Spreading Technique[1]

1. Digestion medium: in 25 ml sterile deionized water, dissolve 0.1 g (final concentration 0.4%) of cytohelicase (Sigma Aldrich, St. Louis, MO, USA), 0.375 g (final concentration 1.5%) of sucrose and 0.25 g (final concentration 1%) of polylvinylpyrrolidone (MW 40,000; Sigma Aldrich, St. Louis, MO, USA). Dispense aliquots of 1 ml and store at –20°C.

2. Spreading medium: dissolve 0.05% Triton X-100 in freshly distilled water.

3. Paraformaldehyde fixative: pre-warm 100 ml of sterile deionized water to 60°C in a microwave and add four drops of 1 M NaOH. Weigh out 4 g paraformaldehyde (EM grade) in the fume hood, add to the pre-warmed water, and stir using a magnetic stirrer until dissolved (this process usually takes around an hour). Filter through Whatman paper, allow to cool, and adjust the pH to 8.0 with 1 M HCl. The fixative can be stored at 4°C for up to 1 week.

4. Blocking buffer: dissolve 1% Bovine Serum Albumin (BSA) in phosphate buffered saline (PBS). We use ready-made tablets to prepare PBS (Oxoid, Fisher Scientific, Suwanee, GA, USA) but it can also be prepared as a 10× stock (1.37 M NaCl, 27 mM KCl, 100 mM Na_2HPO_4, 18 mM KH_2PO_4 pH 7.4) and diluted as necessary. Autoclave before storage at room temperature. Prepare a 1:10 working solution of the blocking buffer in sterile deionized water.

5. Primary antibodies: Make up to the relevant dilution (usually between 1:50 and 1:1,000) of the antibody in PBS buffer containing 1% BSA and 0.1% Triton X-100.

6. Washing solution: PBS buffer containing 0.1% Triton X-100.

7. Secondary antibodies: Anti-rat, anti-rabbit, anti-guinea pig, or anti-mouse (depending on the primary antibody) antibodies conjugated to FITC , Cy-3 (Sigma Aldrich, St. Louis, MO, USA), Texas Red (Vector Laboratories, Burlingame, CA, USA) or Alexa Fluor 350 (Invitrogen, Carlsbad, CA, USA). Make up to the relevant dilution of the antibody (1:50, apart from Texas

---

[1]Adapted with kind permission from Springer Science and Business Media from: Armstrong, S.J., Sanchez –Moran, E., and Franklin, F. C. H. (2009) Cytological analysis of *Arabidopsis thaliana* meiotic chromosomes. *Methods in Molecular Biology 558: Meiosis Cytological Methods*, 131–145.

Red and Cy-3, which should be used at the 1:200 dilution) in PBS buffer containing 1% BSA and 0.1% Triton X-100.

8. DAPI counterstaining solution: prepare a stock solution of 4, 6-diamidino-2-phenylindole (DAPI) at 1 mg/ml in sterile deionized water. Dispense in aliquots and store at –20°C. For use, dilute to 10 µl/ml in an anti-fade mounting medium such as Vectashield (Vector Laboratories, Burlingame, CA, USA).

9. Hot plate set at 33°C.

10. A plastic box containing moist tissue paper to create a humid atmosphere.

---

3   Method

3.1   *Spreading Technique*

1. Collect around 5 buds of approximate length of 0.2–0.4 mm onto damp filter paper. Using a dissecting microscope, quickly remove the anthers from the buds using a mounted needle and fine forceps, yielding a maximum of 30 anthers. Discard any yellow anthers as these will contain pollen. Place around 30 anthers into 10 µl of digestion medium on a clean glass slide and incubate for 4 min at 33°C in a humidified atmosphere (see Note 1).

2. After incubation, tap out individual anthers in the digestion medium with a thin brass rod, to release the pollen mother cells from the anthers. Using a phase-contrast microscope examine the slides to determine if the anthers have been tapped out sufficiently. Early meiocytes tend to stick together: those in leptotene and zygotene can be identified as torpedo shaped columns. Add 10 µl of spreading medium to the suspension and incubate at 33°C in a humidified atmosphere.

3. Using a phase-contrast microscope, monitor the slide carefully for bursting of the cell walls and fix when this occurs (maximum time 4 min).

4. Mark the area of the suspension with a diamond pen and fix with 20 µl paraformaldehyde fixative in a fume hood.

5. When dry, briefly rinse in wash solution and rinse in water.

6. Immerse the slides in wash solution twice for 5 min each wash.

7. Block non-specific binding by incubation with the blocking buffer; 100 µl of the blocking buffer is sufficient. Cover with parafilm and incubate in a humidified atmosphere at room temperature for 45 min.

8. Make up primary antibodies to the relevant dilution. Add 100 µl of the antibody solution and cover with parafilm. Incubate in a humidified atmosphere, either overnight at 4°C or at 37°C for 30 min.

9. Wash twice in the washing solution for 5 min.

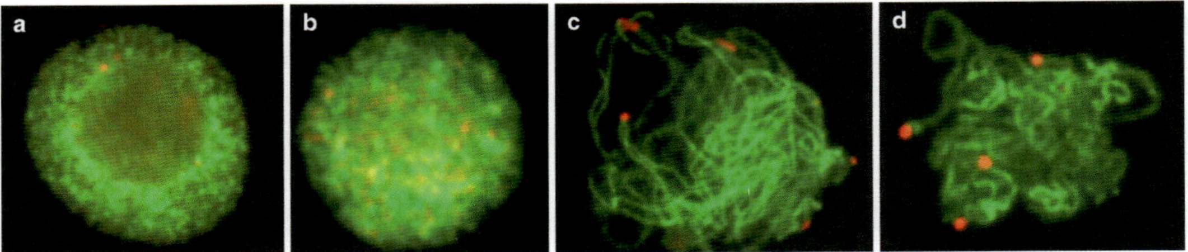

Fig. 1 Dual immunolocalization of AtASY1 and AtRECQ4A (6) in pollen mother cell nuclei of ecotype Columbia (Col) during prophase I (**a–c**). (**a**) G2 chromosomes. Numerous punctate foci of AtASY1 (*green*) and AtRECQ4 (*red*). (**b**) Early leptotene chromosomes showing continuous signals of AtASY1 and numerous small foci of AtRECQ4A. (**c**) Diplotene chromosomes showing considerable enrichment of AtRECQ4A at the telomeres, which is a feature of late prophase (6). (**d**) Dual immunolocalization of AtASY1 and AtRECQ4A in pollen mother cell nucleus at pachytene of ecotype Cape Verde Islands (CVI). Note large signals with AtRECQ4A at the telomeres, most likely due to the significantly longer telomeres in CVI compared to Col

10. Make up the secondary antibody to the relevant dilution. Add 100 μl of the secondary antibody solution and cover with parafilm. Incubate for 30 min in a humidified atmosphere at 37°C in the dark. Keep the preparations in the dark from this stage onwards.

11. Wash twice in the washing solution for 5 min.

12. Optional: immunocytology can be combined with fluorescence in situ hybridization (FISH) (see Note 2) or with 5-bromo-2'-deoxyuridine (BrdU) labelling (see Note 3).

13. Mount the preparation in 7 μl of DAPI counterstaining solution and view with a fluorescence microscope equipped with an image capture and analysis system. Figure 1 shows an example of dual immunolocalization using this technique.

4 Notes

1. Alternatively, rather than dissecting out the anthers on damp filter paper, they can be removed from individual buds that have been placed directly into the enzyme mix. The advantage of this procedure is that it avoids the risk of drying out of the anthers. On the other hand, anthers must be dissected out within 4 min. Continue with the protocol from step 2 onwards.

2. It is possible to combine immunolocalization with FISH. Make slides for the combined technique using the spreading technique to the end of Subheading 3.1, step 9. Following washing, apply an anti-biotin antibody (raised in rabbit or rat, depending on the primary antibody) at a concentration of 1:50 in immunolocalization blocking buffer. This procedure protects the primary antibody in the subsequent FISH procedure.

Incubate for 30 min at 37°C in a humidified atmosphere. Following washing, take the slides through an alcohol series: immerse sequentially in 70, 85, and 100% ethanol for two minutes each. The slides can then be used for the FISH procedure (see Chapter 2, Subheading 3.2). Note that either directly labelled probes or digoxigenin-labelled FISH probes must be used due to the use of biotin in the immunolocalization protocol. Following the last wash in the FISH protocol, apply an anti-biotin FITC or Cy3 secondary antibody to detect the biotinylated primary antibody. Incubate for 30 min at 37°C in a humidified atmosphere. Wash as before and mount in DAPI counterstaining solution.

3. It may be useful to combine the immunolocalization technique with a time course experiment in order to assess the relationship between protein localization and the duration of specific meiotic stages. This analysis may be carried out using BrdU pulse-labelling as described in the time course protocol (see Chapter 12, Subheading 3.1). However, slides should be prepared using fresh material, as described in this chapter, rather than fixed material. The antibody should be detected first, followed by detection of BrdU.

Acknowledgments

The research leading to these results has received funding from the European Community's Seventh Framework Programme FP7/2007–2013 under grant agreement number KBBE-2009-222883. We also acknowledge support by the Royal Society and the Biotechnology and Biological Sciences Research Council (BBSRC). Technical support has been provided by Steve Price.

References

1. Jones GH, Armstrong SJ, Caryl AP, Franklin FCH (2003) Meiotic chromosome synapsis and recombination in *Arabidopsis thaliana*; an integration of cytological and molecular approaches. Chromosome Res 11:205–215

2. Armstrong SJ, Caryl AP, Jones GH, Franklin FCH (2002) Asy1, a protein required for meiotic chromosome synapsis, localizes to axis associated chromatin in *Arabidopsis* and *Brassica*. J Cell Sci 115:3645–3655

3. Higgins JD, Armstrong SJ, Franklin FCH, Jones GH (2004) The Arabidopsis *MutS* homologue *AtMSH4* functions at an early step in recombination; evidence for two classes of recombination in *Arabidopsis*. Genes Dev 18:2557–2570

4. Higgins JD, Sanchez-Moran E, Armstrong SJ, Jones GH, Franklin FCH (2005) The *Arabidopsis* synaptonemal complex protein ZYP1 is required for chromosome synapsis and normal fidelity of crossing over. Genes Dev 19:2488–2500

5. Albini SM, Jones GH, Wallace BMN (1994) A method for preparing two–dimensional surface-spreads of synaptonemal complexes from plant meiocytes for light and electron microscopy. Exp Cell Res 152:280–285

6. Higgins JD, Ferdous M, Osman K, Franklin FCH (2011) The RECQ helicase AtRECQ4A is required to remove inter-chromosomal telomeric connections that arise during meiotic recombination in *Arabidopsis*. Plant J 65:492–502

Immunolocalization Protocols for Visualizing Meiotic Proteins in *Arabidopsis thaliana*: Method 3

Xiaohui Yang, Li Yuan, and Christopher A. Makaroff

Abstract

Immunolocalization studies to visualize the distribution of proteins on meiotic chromosomes have become an integral part of studies on meiosis in the model organism *Arabidopsis thaliana*. This chapter describes fixation, sample preparation, and immunolocalization techniques used in our laboratory. These techniques have been used to visualize a wide range of meiotic proteins involved in different aspects of meiosis, including sister chromatid cohesion, recombination, synapsis, and chromosome segregation.

Keywords *Arabidopsis*, Immunofluorescence, Meiosis

1 Introduction

Arabidopsis thaliana initially emerged as a model organism to study meiosis because of its molecular and genetic attributes, including the small size of its genome. This is in contrast to the large chromosomes of most flowering plants, which are ideal for cytological studies on meiosis, but are much more difficult to analyze at the molecular level. However, the relatively small size of the *Arabidopsis* chromosomes also made cytological studies more difficult. Over the past decade advances in microscopy and technique development have overcome many of the problems associated with doing high-resolution cytology on *Arabidopsis* meiocytes. These developments coupled with the molecular strengths of *Arabidopsis* have resulted in a wealth of information on meiosis and the roles of meiotic proteins in plants (1). At the core of these cytological studies has been the development of various techniques that allow the visualization of proteins on *Arabidopsis* chromosomes.

A number of different fixation and spreading techniques have been developed to prepare *Arabidopsis* meiotic chromosomes for immunolocalization studies (2–6). While all of these techniques have

Wojciech P. Pawlowski et al. (eds.), *Plant Meiosis: Methods and Protocols*, Methods in Molecular Biology, vol. 990, DOI 10.1007/978-1-62703-333-6_11, © Springer Science+Business Media New York 2013

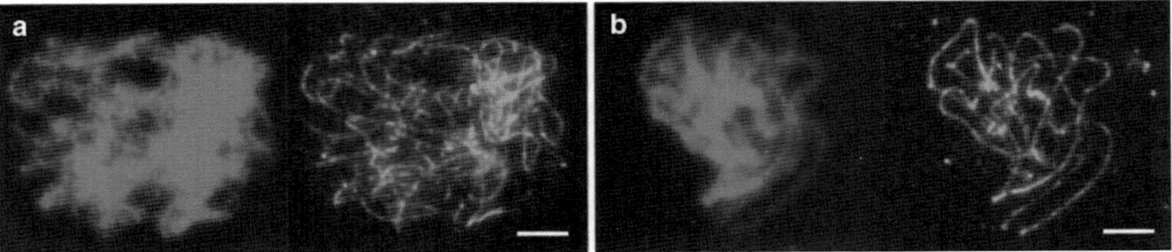

Fig. 1 Immunolocalization of SMC3 in paraformaldehyde-fixed male meiocytes. (**a**) Zygotene. (**b**) Pachytene. Chromosomes are counterstained with DAPI (*left image of each panel*); SMC3 (*right image of each panel*). Bar = 10 μm

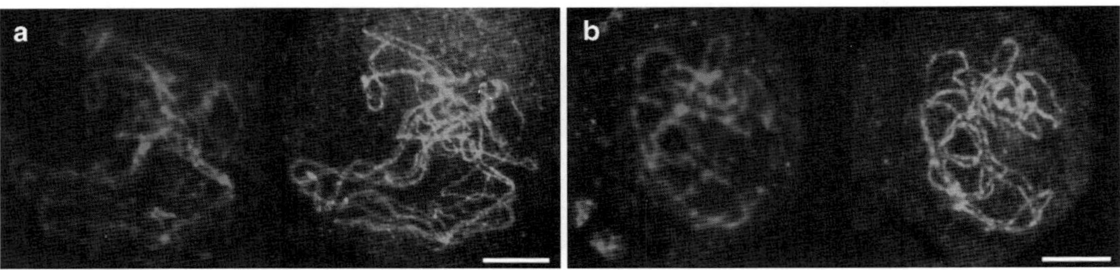

Fig. 2 Immunolocalization of SMC3 in methanol/acetone-fixed male meiocytes. (**a**) Zygotene. (**b**) Pachytene. Chromosomes are counterstained with DAPI (*left image of each panel*); SMC3 signal is (*right image of each panel*). Bar = 10 μm

been shown to produce high quality results, the methods have different strengths and weaknesses and certain methods are better than others depending on the question being investigated and the desired outcome. The two techniques most often used in our laboratory for sample fixation prior to immunolocalization utilize either paraformaldehyde (Fig. 1) or methanol/acetone (Fig. 2). Paraformaldehyde is the most commonly used fixative to prepare samples for meiotic spreads in *Arabidopsis*. Paraformaldehyde fixation results in intermolecular cross-linking of proteins through free amino groups in the tissue samples, resulting in strong preservation of the cellular and chromosome structure. Therefore, additional cell permeabilization steps, such as flash-freezing, squashing by hard physical force, enzyme digestion, and detergent treatment are typically used to provide for better access to chromosomal proteins during immunolocalization on paraformaldehyde fixed meiocytes.

Methanol/acetone fixation extracts the cellular lipids and dehydrates cells quickly, disrupting hydrophobic interactions and resulting in the precipitation and aggregation of proteins. Methanol/acetone fixation still preserves the cellular architecture, but can result in reduced antigenicity for some proteins. However, it can also allow better access of the antibody to some antigens depending on their position in a complex structure. Perhaps most important, extensive cellular permeabilization steps are typically not needed after methanol/acetone fixation. Therefore, immunolocalization procedures utilizing methanol/acetone fixation tend to be shorter than those using paraformaldehyde fixation. It is also useful if one is interested in maintaining the integrity of the nucleoplasm and cytoplasm. For example, methanol/acetone fixation was used to show that *Arabidopsis* SMC3 localizes to both the chromosomes and the meiotic spindle (3).

This chapter describes immunolocalization protocols commonly used in our laboratory for the analysis of protein distribution during meiosis in *Arabidopsis* anthers. Protocols for both methanol/acetone and paraformaldehyde fixation followed by immunolocalization are presented. No matter which method is used, some optimization of the procedures will be necessary depending on the antigen, its subcellular environment, and the antibody preparation that is being used.

2   Materials

2.1   Plants

Five to ten *Arabidopsis* seeds should be sown in 4 in. pots containing a commercial soil mix, such as Metro-Mix360 (SunGro Horticulture, Vancouver, BC, Canada). Grow plants in a plant growth chamber at 23°C with a 16 h light/8 h dark cycle under fluorescent illumination at an intensity of 200 mE/m²s. Growth of plants at a relatively low density with appropriate amounts of water is important to obtain healthy plants and the best results (see Note 1).

2.2   Tools and Equipment

1. Dry ice or –80°C freezer

2. Coverslips and poly-L-lysine or 3-methacryloxypropyl-trimethoxysilane coated slides: dip the precleaned slides (Corning Inc., Corning, NY, USA) into the slide coating solution (see Subheading 2.3), remove, drain, and dry at room temperature for at least 2 h or on a plate warmer (42–55°C) for 30 min. Wipe-clean with paper towel wet by 95% ethanol.

3. Metal agarose/gelatin gel applicator (Fig. 3).

4. Slide incubation box with two parallel rails for supporting slides and water-wet paper towels for maintaining humidity levels.

5. Dissecting needles or syringe with needles (0.5–1 ml).

6. Forceps with long, fine tips.

7. Commercially available water repellent slides such as the *Lab-Tek™* Chamber Slide™ System (Thermo Fisher Scientific, Waltham, MA, USA) (see Note 2).

8. Stereomicroscope.

9. Florescence microscope and image acquisition software.

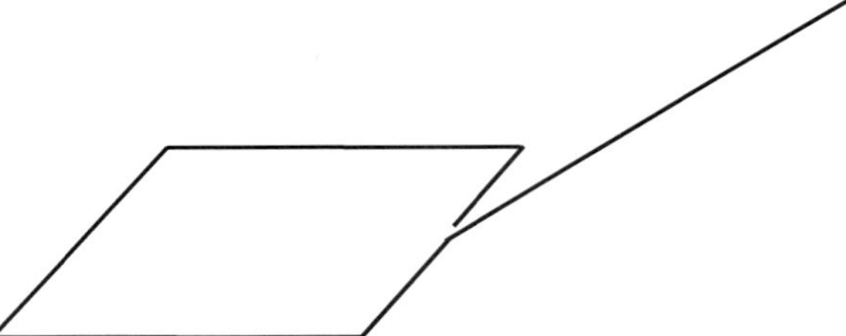

Fig. 3 The agarose/gelatin gel applicator. The applicator is made of thin metal wire (2.0 × 1.2 in.)

2.3 Solutions

1. Slide coating solution: add 50 μl of 0.1% (w/v) poly-L-lysine (Sigma Aldrich, St. Louis, MO, USA) to 50 ml of deionized water. Alternatively, use a mix of 20 μl of 3-methacryloxypropyl-trimethoxysilane in 50 ml of 95% ethanol, 0.5% glacial acid.

2. Agarose/gelatin solution: melt 0.94 g of low melting agarose (Fisher Scientific, Suwanee, GA, USA), 0.84 g of gelatin (Sigma Aldrich, St. Louis, MO, USA), and 2.5 g of sucrose in 100 ml of deionized water (see Note 3).

3. Agarose plates for bud dissection: melt 0.8 g of agarose (Fisher Scientific, Suwanee, GA, USA) in 100 ml of deionized water. Pour melted agarose solution into plastic Petri dishes. The plates can be kept at 4°C for a few weeks.

4. 1× PBS buffer: dissolve 8 g of NaCl, 0.2 g of KCl, 1.44 g of Na_2HPO_4, 0.24 g of KH_2PO_4 in 800 ml of deionized H_2O. Adjust the pH to 7.4 and the volume to 1 l with additional deionized H_2O. Sterilize by autoclaving.

5. Methanol/acetone fixative solution: mix methanol and acetone at 4:1 (v/v) ratio and leave on ice. Prepare fresh just before use.

6. Buffer A: 15 mM PIPES, 80 mM KCl, 20 mM NaCl, 2 mM EDTA, 0.5 mM EGTA, 0.5 mM spermidine, 0.15 mM spermine, 1 mM DTT, adjust pH to 5.8 (7).

7. Paraformaldehyde (4%) fixative in buffer A: 4 g of paraformaldehyde in 100 ml of buffer A containing 0.01% Triton X-100. Heat at 55°C for up to 2 h with occasional mixing or until the solution becomes clear.

8. Toluidine Blue solution (0.5%): dissolve 0.5 g of solid Toluidine Blue in 100 ml of deionized water.

9. 2 N hydrochloric acid (HCl): dilute 100 μl of concentrated HCl (12 N) in 500 μl of deionized water. Prepare fresh each time.

10. Citrate buffer (10 mM, pH 4.5): mix 4.45 ml of 0.1 M sodium citrate with 5.55 ml of 0.1 M citric acid. Make up the volume to 100 ml with deionized water.

11. Enzyme digestion solution: dissolve 3 mg of pectolyase Y23 (Seishin Corporation, Kobe, Japan), 3 mg of cellulose (Sigma Aldrich, St. Louis, MO, USA), 15 mg of β-glucuronidase (Sigma Aldrich, St. Louis, MO, USA), and 15 mg of sucrose in 1 ml of 10 mM Citrate Buffer pH 4.5. Aliquot and store at -20°C.

12. Wash buffer: 1× PBS, 0.1% Tween 20, 1 mM EDTA.

13. Cell wall permeabilization solution: mix 100 μl of Triton X-100 in 10 ml in washing buffer.

14. 2% blocking buffer: dissolve 0.2 g BSA in 10 ml of washing buffer.

15. Chromosome counter-stain solution: 4, 6-diaminido-2-phenlyinidole (DAPI) in Vectashield (Vector Laboratory, Burlingame, CA, USA).

3   Methods

3.1   Bud Preparation

1. Remove an inflorescence from a healthy plant when the first bud blooms and place it on an agarose plate (see Note 4).

2. Open the buds under stereomicroscope and remove any buds with yellow anthers that contain pollen.

3. Dissect out one of the 6 anthers from the next largest bud.

4. Crush the anther in a drop of toluidine blue staining solution under a coverslip.

5. Examine the anther contents using a light microscope. Discard buds that contain pollen or microspores.

6. Keep examining anthers until buds containing meiocytes at tetrad stage are seen (see Note 5).

7. Fix the remaining buds in the inflorescence in the desired fixative (see Subheadings 3.2 and 3.3).

3.2   Method 1: Methanol/Acetone Fixation

1. Fix approximately 10 inflorescences in 500 μl of methanol/acetone for up to 2 h on ice or at 4°C. Change the buffer a few times until the buds become whitish.

2. Wash the buds three times with 1 ml of deionized water, 5 min each wash.

3. Dissect anthers out of the buds from 3 to 5 inflorescences in water on a coated slide (see Subheading 2.2 and Note 2). Place anthers at one end of the slide in a small amount of water. Add water as needed to prevent the anthers from drying.

4. Remove as much water as possible with a fine syringe needle and add 20 μl of 2 N HCl. Incubate for 3–5 min at room temperature. Longer acid treatment will prevent chromosomes from staining by DAPI.

5. Remove all of the HCl solution by a fine syringe needle and immediately add 50 µl of sterile deionized water and leave for 10 min at room temperature. Most anthers will become semi-transparent after the incubation.

6. Remove most of the water by a fine syringe needle, leaving approximately 3–5 µl of water to keep the anthers wet.

7. Macerate the anthers with fine forceps into small pieces. Prevent the sample from drying by adding 1–2 µl of water while breaking up the anthers (see Note 6).

8. Add an additional 10–15 µl of deionized water depending on the number of anthers.

9. Transfer equal amounts of the broken tissue onto three newly coated slides using fine forceps (see Note 7).

10. Spread the sample with forceps over an area slightly smaller than a coverslip. Cover the sample with a coverslip with one corner of the coverslip extending over the edge of the slide for easy removal later.

11. Sandwich the slide in a folded paper towel with the flat edge of the slide against the folded side of the paper towel to prevent the coverslip from moving. Fold the paper towel over the top of the coverslip and hold firmly (see Note 8).

12. Squash the sample area by gentle tapping a few times with a pencil rubber eraser (Important: too much force will break the chromosomes).

13. Remove the slide from the paper towel sandwich. Freeze the slide in a –80°C freezer or on dry ice for 15 min.

14. Remove the coverslip while frozen and dry the slide up to overnight at room temperature. The slide can be used right away for immunolocalization or stored at –80°C for future use.

3.3 Method 2: Paraformaldehyde Fixation

1. Stage and remove old buds from the inflorescence (see Subheading 3.1).

2. Dissect 100–120 anthers (more anthers will increase the yield of meiocytes) out of buds from three to five inflorescences on 0.8% agarose plate. The dissected anthers can be kept on the agarose plate prior to fixation.

3. Transfer 50 µl of 4% paraformaldehyde fixative solution onto a *Lab-Tek* Chamber Slide.

4. Transfer all anthers into the fixative solution on the slide.

5. Fix the anthers for 60–80 min on ice (see Note 9).

6. Wash the anthers three times, 5 min each, with Buffer A.

7. Remove the wash buffer and add 10–20 µl of water to the anthers. Cover the anthers with a coverslip (Fig. 4, see Note 10).

8. Under a dissecting microscope, use forceps to gently press down on the coverslip to release the meiocytes from the

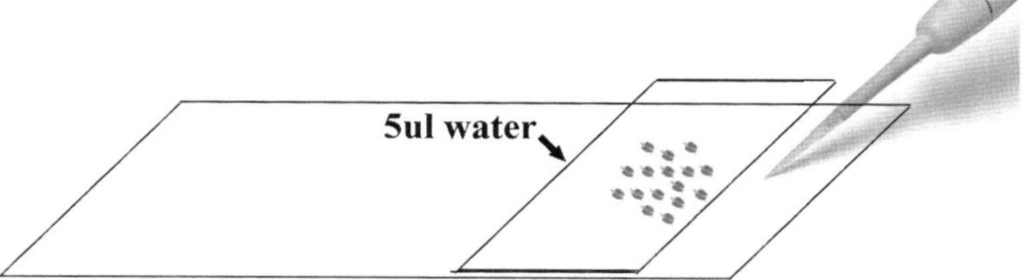

Fig. 4 Diagram of collecting meiocytes from slide

anthers. Meiocytes are released as an associated cluster of cells or as individual cells that are visible under the microscope.

9. Add 5–10 µl of water to the left side of the coverslip and use a pipette with a fine tip to collect the solution that contains meiocytes from the other side of the coverslip (Fig. 4).

10. Transfer the meiocytes onto a new coated slide. Repeat steps 3 and 4 until meiocytes are no longer released from squashing of the anthers. All of the collected meiocytes should be together (see Note 11).

11. Put the slide with collected the meiocytes on dry ice for at least 10 min (see Note 12).

12. Transfer the slide to room temperature and let the meiocytes thaw.

13. Mix and transfer 5 µl of the meiocyte solution onto each of 5–10 new coated slides and cover the cells with a coverslip.

14. Put the slides in a folded paper towel (see Subheading 3.2, steps 10 and 11). Squash the sample area by hard tapping a few times with a pencil rubber eraser (see Note 8).

15. Freeze the slide in a –80°C freezer or on dry ice for 15 min.

16. Remove the coverslip while still frozen and dry the slides up to overnight at room temperature. The slides can be used immediately or stored at –80°C for future use.

3.4 Cell Permeabilization (See Note 14)

1. Melt agarose/gelatin solution and apply a thin layer of the gel over the dried sample area using the agarose/gelatin applicator. Dry on warm plate (no more than 50°C) (see Note 13).

2. Soak the sample area with 500 µl of wash buffer for 30–60 min to rehydrate the cells.

3. Drain the wash buffer by leaning the slides on a paper towel. Add 50 µl of enzyme digestion solution and cover the area with parafilm (see Note 14). Do not use a glass coverslip as it will damage the membrane when the coverslip is removed.

4. Incubate the slide in a slide incubation box at 37°C for 60–90 min.

5. Before removing the parafilm, add from the side 500 μl of ice-cold wash buffer for 5 min to cool off the agarose-gelatin membrane.

6. Carefully remove the parafilm. Drain the wash solution and replenish with 500 μl of cell permeabilization solution at room temperature.

7. Incubate the slide with slow agitation in an incubation box for 60 min at room temperature.

8. Wash slides twice with 500 μl of wash buffer for 5 min each wash.

3.5 Immuno-localization Procedure

1. Block the sample in 200 μl of blocking buffer for 1 h. If cell permeabilization (Subheading 3.4) is not used, then the sample must be rehydrated as in Subheading 3.4, step 2.

2. Dilute primary antibody (100- to 1,000-fold) in blocking buffer.

3. Apply 100 μl of diluted antibody to the sample area and cover with parafilm.

4. Incubate the slides at 37°C for at least 1 hr or at 4°C overnight in a slide incubation box (see Note 15).

5. The next day wash out the antibody solution with 1–2 ml of wash buffer. Wash the slides with 500 μl of wash buffer for five times, 20 min each, with slow agitation in a slide incubation box. Add more buffer if needed.

6. Dilute the secondary antibody (300- to 1,000-fold) in blocking solution.

7. Remove the wash buffer, apply 100 μl of secondary antibody to the sample area and cover the slide with parafilm.

8. Incubate at 37°C for at least 1 h or overnight at 4°C in the slide incubation box (see Note 15)

9. Wash out the secondary antibody solution with 1–2 ml of wash buffer. Wash the slide with 500 μl wash buffer five times, 20 min each, with slow agitation in a slide incubation box. Add more buffer if needed.

10. Remove excess wash buffer and apply 5 μl of DAPI in Vectarshield. Cover the slide with a coverslip. Examine slides with a fluorescence microscope equipped with a digital image system and image acquisition software.

4 Notes

1. Strong, healthy plants should only be used for immunolocalization, if possible. Healthy plants usually produce more than ten floral buds in one inflorescence, which will contain meiocytes at all meiotic stages. More inflorescences should be col-

lected if the plants are weak. Therefore, plants should not be over crowded. Proper watering is also important for the growth of healthy plants.

2. *Lab-Tek* Chamber Slides (Thermo Fisher Scientific, Waltham, MA, USA) are recommended for anther dissection from fixed buds and incubation because the slides can be washed and reused without losing the water repellence. *Lab-Tek* Chamber Slides hold the water very well without dispersing. The slides also have a cover to prevent water evaporation. The chamber should be removed before use. Alternatively, pre-cleaned slides can be coated with the water repellent 3-methacryloxypropyl-trimethoxysilane.

3. The gel mix can be kept at 4°C for several weeks. Discard if contaminated with bacteria.

4. Agarose plates can be used for dissecting anthers from fresh buds. The plate can prevent the buds and dissected anthers from drying out and can be used for short term storage of floral buds at 4°C.

5. Pre-staging the anthers prevents the contamination of samples with large amounts of pollen that will decrease the efficiency of the enzyme digestion and immune reaction. With some experience this step can be skipped and floral buds can be collected based on their sizes.

6. Broken anthers should not be allowed to dry out. However, too much water will make it hard to break up the anthers. Add only minimum (1 to 2 µl) sterile deionized water when needed.

7. An additional 2 to 3 µl of sterile deionized water can be added to the transferred samples on a coated slide for better spreading with forceps. White or yellow pipette tips can be used for transferring the cell solution. However, use with care as the tissue/cells may stick to the tip wall.

8. This step is critical. Cell distortion will occur if there is any movement between the slide and the coverslip. Avoid any movement by slowly and firmly pressing the paper towel on the coverslip with two fingers before tapping. For methanol/acetone fixed cells treated with HCl, gentle tapping of the sample area is enough. Excessive use of force will alter chromosome structure or even break the chromosomes. For paraformaldehyde fixed cells, hard tapping is needed to break the meiotic cell wall to release the chromosomes. Optimization of this step may be needed to get the best chromosome morphology.

9. Paraformaldehyde fixation is a stronger fixative which results in better resolution of the chromosomes. However, longer fixation will make it difficult to break the cell wall and hard for the antibody to enter the meiocytes. Therefore, optimization of the fixing time may be needed.

10. Collection of meiocytes is more efficient when using half of a coverslip. However, a whole coverslip can also be used at this step when there are more anthers. It may be necessary to add more water to the anthers releasing the meiocytes.

11. The meiocytes collected by this method contain less cell debris from the anther wall. It will be more efficient to obtain the meiotic chromosomes at later steps. While pressing, make sure not to break the coverslip, and to keep the anthers within the coverslip at all times.

12. Quick-freezing of anthers on dry ice is necessary to release the chromosomes more efficiently. This process will soften the cell wall and break the cell membrane.

13. Use of a gel overlay is optional. However, gel will help to prevent the sample from drying out during the experiment. It will also help to achieve more uniform signals.

14. Enzyme digestion and the use of a cell wall permeabilization solution will help the penetration of antibody into the cells to obtain better signals.

15. Incubation at 4°C overnight is recommended to reduce background and enhance signal.

Acknowledgment

Research on meiosis in the Makaroff lab has been supported by a grant (MCB0718191) from the NSF.

References

1. Chang F, Wang YX, Wang SS, Ma H (2011) Molecular control of microsporogenesis in *Arabidopsis*. Curr Opin Plant Biol 14:66–73

2. Cai X, Dong FG, Edelmann RE, Makaroff CA (2003) The *Arabidopsis* SYN1 cohesin protein is required for sister chromatid arm cohesion and homologous chromosome pairing. J Cell Sci 116:2999–3007

3. Lam WS, Yang XH, Makaroff CA (2005) Characterization of *Arabidopsis thaliana* SMC1 and SMC3: evidence that ASMC3 may function beyond chromosome cohesion. J Cell Sci 118: 3037–3048

4. Armstrong S, Caryl A, Jones G, Franklin F (2002) ASY1, a protein required for meiotic chromosome synpasis, localizes to axis-associated chromatin in *Arabidopsis* and *Brassica*. J Cell Sci 115:3645–3655

5. Armstrong SJ, Franklin FCH, Jones GH (2003) A meiotic time-course for *Arabidopsis thaliana*. Sex Plant Reprod 16:141–149

6. Chelysheva L, Grandont L, Vrielynck N, le Guin S, Mercier R, Grelon M (2010) An easy protocol for studying chromatin and recombination protein dynamics during *Arabidopsis thaliana* meiosis: immunodetection of cohesins, histones and MLH1. Cytogenet Genome Res 129: 143–153

7. Dernburg AF, Sedat JW, Hawley RS (1996) Direct evidence of a role for heterochromatin in meiotic chromosome segregation. Cell 86: 135–146

Chapter 12

A Time Course for the Analysis of Meiotic Progression in *Arabidopsis thaliana*

Susan Armstrong

Abstract

We have developed a technique that enables the thymidine analogues bromodeoxyuridine (BrdU) or ethynyldeoxyuridine (EdU) to be taken up via the transpiration stream into pollen mother cells and incorporated into the meiotic S-phase. Labelled inflorescences are sampled at intervals, which enables assessment of the progression of labelled meiocytes by detection of the incorporated BrdU or EdU.

Keywords *Arabidopsis*, Meiosis, Time course, BrdU, EdU

1 Introduction

In order to gain an understanding of the cytological and molecular events occurring in *Arabidopsis* meiosis, they need to be placed in a temporal framework of the meiotic pathway. This approach has been instrumental in linking key molecular events with chromosome and nuclear changes in yeast (1–3). During the 1970s, the duration of meiosis in a few plant species was investigated and relied either on sampling single anthers from inflorescences and subsequently taking the remaining anthers after timed intervals or by introducing tritiated thymidine to the meiotic S-phase. These methods have a number of disadvantages, for example using the former method is likely to compromise living material and the latter method is time consuming (4 *and references therein*). More recently these methods have been supplanted by the use of the thymidine analogues, bromodeoxyuridine (BrdU) and ethynyldeoxyuridine (EdU) both of which can be detected immunologically.

We have found that we can introduce these analogues into the transpiration stream of *Arabidopsis* and have observed that they are incorporated both into the meiotic and mitotic S-phase. We have

Wojciech P. Pawlowski et al. (eds.), *Plant Meiosis: Methods and Protocols*, Methods in Molecular Biology, vol. 990,
DOI 10.1007/978-1-62703-333-6_12, © Springer Science+Business Media New York 2013

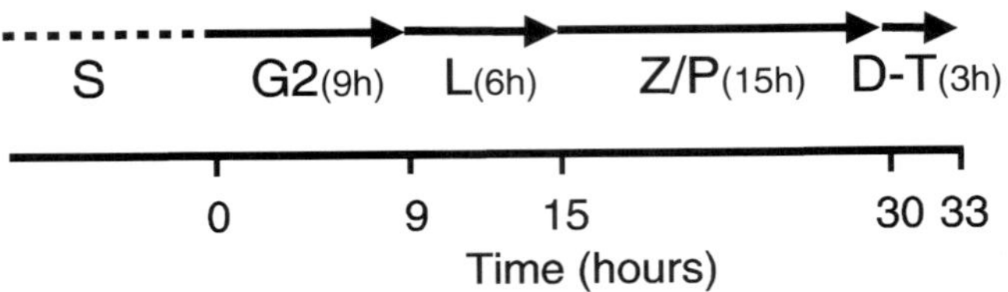

Fig. 1 Duration of meiotic stages in *Arabidopsis thaliana*. The diagram summarizes the durations of the meiotic stages in a typical *Arabidopsis* (ecotype Columbia) time course for the progression of the pollen mother cells through meiosis. *S* meiotic S-phase, *L* leptotene, *Z/P* zygotene/pachytene, *D-T* diplotene to tetrads

produced a temporal framework that will form a basis for further analyses of the timing and relationships of events within the *Arabidopsis* meiotic pathway (Fig. 1).

2   Materials

2.1   Plant Material

Sow seeds of *Arabidopsis* in 6 cm diameter pots in soil-based compost. Grow the plants in dedicated growth chambers maintained at 18°C with a 16 h light cycle. We sow around 25 seeds as this will give us sufficient plants for duplicates at each time point as well as some reserve plants. The plants we use are those with terminal inflorescences in the range of 6–8 cm height.

2.2   Marking the Meiocyte S-Phase

1. BrdU, available as a 10 mM solution in the 5-bromo-2′-deoxyuridine Labelling and Detection Kit I (Roche Applied Science, Indianapolis, IN, USA) (see Note 1).
2. EdU, available as a 10 mM solution in the Click-iT EdU Alexa Fluor 594 HCS Assay Kit (Invitrogen, Carlsbad, CA, USA) (see Note 2).

2.3   Fixation and Preparation of Slides

1. Fixative: mix 3 parts of absolute ethanol with 1 part of glacial acetic acid and keep on ice. Prepare fresh fixative as required and discard at the end of the day.
2. 0.01 M Citrate Buffer: prepare a working solution of the buffer (pH 4.5) by using 4.45 ml of 0.1 M sodium citrate, 5.55 ml of 0.1 M citric acid, made up to 100 ml with sterile deionized water.
3. Stock digestion medium: dissolve 1% cellulase (Sigma Aldrich, St. Louis, MO, USA), 1% pectolyase (Sigma Aldrich, St. Louis, MO, USA) in a working solution of the 0.01 M citrate buffer, pH 4.5, store in aliquots at –20°C.
4. Digestion medium: mix 333 μl of the stock digestion medium with 667 μl of 0.01 M citrate buffer, pH 4.5.

5. 60% acetic acid: dilute glacial acetic acid with sterile deionized water.

6. 5-Bromo-2′-deoxy-uridine Labelling and Detection Kit I (Roche Applied Science, Indianapolis, IN, USA).

7. Click-iT EdU Alexa Fluor 594 HCS Assay Kit (Invitrogen, Carlsbad, CA, USA)

8. DAPI: prepare a stock solution of 4′,6-diamidino-2-phenylindole (DAPI) in sterile deionized water at 1 mg/ml, dispense in aliquots and store at –20°C. For use, add 10 μl of this stock to 1 ml of an anti-fade mounting medium, Vectashield (Vector Laboratories, Burlingame, CA) and store at 4°C.

3   Methods

3.1   Marking the Meiocyte S-Phase

1. Cut under water stems (6–8 cm tall) from well-grown *Arabidopsis* plants with terminal inflorescences with only one or two open flowers. We generally use 20 stems per time course experiment (see Note 3).

2. The cut stems are quickly transferred to BrdU solution or EdU solution for 2 h for their uptake by the transpiration stream and their incorporation into cells in S-phase.

3. After 2 h, remove the stems from the BrdU or EdU solution. Rinse the ends of the stems in tap water and place the stems in small glass bottles containing tap water. Maintain the labelled cut stems in a controlled environment glasshouse at 18°C (range 17–23°C)

3.2   Sampling Time Points

1. Sample and fix duplicate inflorescences immediately before the pulse (control), at the end of the 2 h pulse (time point 0 h) and then at 2- or 4-h intervals up to 40 h (see Note 3).

2. To fix, cut terminal inflorescences with around 1 cm of stem and immediately place them in freshly made ice-cold fixative. Leave them on the bench at room temperature and replace the fixative after 2–3 h. The fixed material is suitable for digestion 24 h after fixation, and can be stored at –20°C for up to 6 months.

3.3   Fixation and Preparation of Slides

Preparation of the slides is similar to this described in the protocol in Chapter 1.

1. Place single, fixed inflorescences in watch glasses (preferably colored black) with fresh fixative. Using a mounted needle and fine forceps (e.g., watchmaker's forceps), divide up the inflorescence into individual buds. Discard buds with yellow anthers (these contain pollen and be post-meiotic), and any larger buds. Retain the remaining buds.

2. Replace the fixative with the working citrate buffer (2×5 min). After that time, replace the citrate buffer with the working solution of the enzymes for 75–90 min (shorter times are better to preserve the organization of the meiocytes in early meiosis). Incubate the buds in a humidified atmosphere (e.g., a sandwich box containing damp tissues) at 37°C for 30 min. Remove the enzyme solution and replace it with the working citrate buffer at 4°C in order to stop the reaction.

3. Place a single bud onto a slide, with a minimum of buffer. Macerate the bud quickly with a mounted needle, ensuring that the material does not dry out.

4. Add 10 μl of 60% acetic acid, mix with the material on the slide, and place the slide on a hot block at 45°C for up to 30 s. Mark the region of the slide containing the material with a diamond pen.

5. Place the slide on the bench and add 100 μl of ice-cold fixative as a ring around the material. Dry the preparation. We find using a commercial hair drier suitable for this purpose.

6. Detection of BrdU/EdU: follow manufacturer's instructions in the kits.

7. Mount the slides in 7 μl DAPI in Vectashield.

8. View the slides with a fluorescence microscope and image captured with an image analysis system (see Fig. 2 for typical cells labelled with EdU or BrdU).

3.4 Analysis of Time Course

For analysis, we score each slide for the time of the first appearance of the BrdU/EdU label at progressively later defined stages, e.g., meiotic interphase, leptotene, zygotene/pachytene, diplotene, diakinesis/metaphase I, metaphase II/tetrad, at each sampling time (4).

---

4 Notes

1. BrdU can also be ordered separately if required (Sigma Aldrich, St. Louis, MO, USA). Make up to 10 mM in sterile 1× PBS and store at –20°C.

2. EdU can be ordered separately as larger volumes may be required (Invitrogen, Carlsbad, CA, USA). Make up to 10 mM in sterile 1× PBS and store at –20°C.

3. Sampling times: samples at 4-h intervals are sufficient to get an overview of the meiotic time course. For a more detailed analysis of events during prophase, 2-h intervals should be used and, consequently, more plants need to be grown.

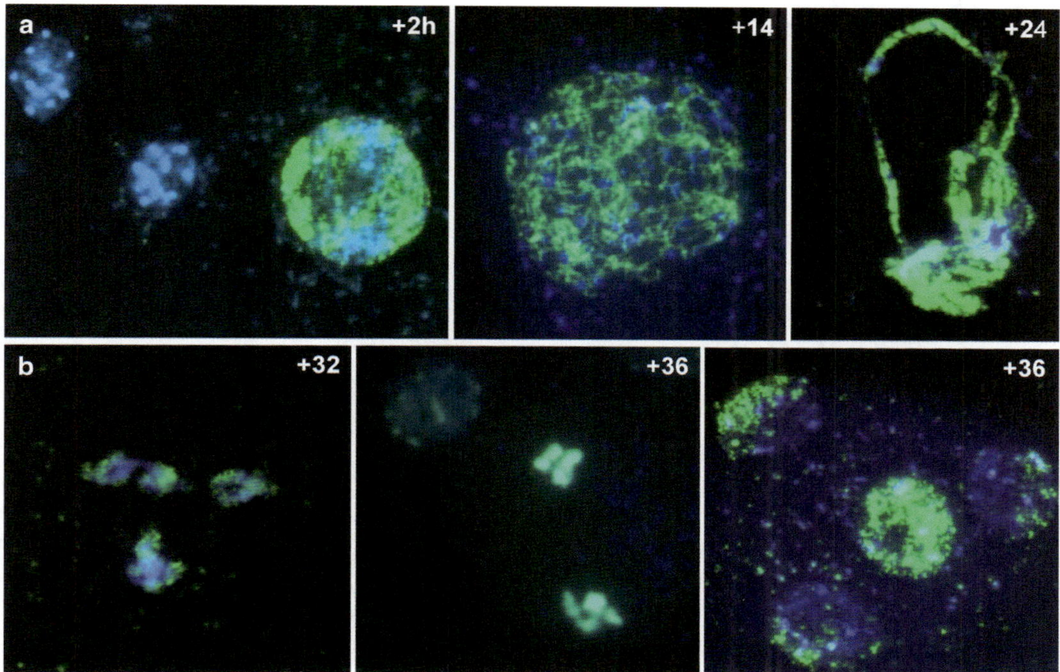

Fig. 2 Detection of EdU/BrdU labelling of *Arabidopsis* pollen mother cells. (**a**) Pollen mother cells sampled at +2 h in the pre-meiotic S phase, +14 h in leptotene and at +24 h in pachytene labelled with EdU. (**b**) Pollen mother cells sampled at +32 h in diakinesis, +36 h in metaphase II (*left*) and tetrads (*right*) labelled with BrdU

Acknowledgments

The work in our laboratories is supported by the Biotechnology and Biological Sciences Research Council (BBSRC) and the European Commission, FP7 grant 222882. Technical support has been provided by Karen Staples and Steve Price.

References

1. Padmore R, Cao L, Kleckner N (1991) Temporal comparison of recombination and synaptonemal complex formation during meiosis in *S. cerevisiae*. Cell 66:1239–1256

2. Hunter N, Kleckner N (2001) The single–end invasion; an asymmetric intermediate at the double-strand break to double-Holliday junction of meiotic recombination. Cell 109: 59–70

3. Terasawa M, Ogawa H, Tsukamotom Y, Shinohara M, Shirahige K, Kleckner N et al (2007) Meiotic recombination–related DNA synthesis and its implications for cross-over and non-cross-over recombinant formation. PNAS 104:5965–5970

4. Armstrong SJ, Franklin FCH, Jones GH (2003) A meiotic time-course for *Arabidopsis thaliana*. Sex Plant Reprod 16:141–149

Chapter 13

Analyzing Meiotic Chromosomes in Rice

Zhukuan Cheng

Abstract

Development of techniques to analyze pachytene chromosomes has greatly overcome most of the difficulties in cytological studies of rice chromosomes caused by their small size. Visualization of meiotic chromosomes has now become routine in cytogenetic studies in this species. This chapter provides protocols on basic meiotic chromosome preparation, FISH analysis, and immunocytology in rice.

Keywords Rice, Meiosis, Chromosomes, FISH, Immunofluorescence

1 Introduction

Besides being the staple food of nearly half of mankind, rice is also one of the most frequently used model plant species for genetic and molecular biology studies (1). The 430 Mb-in size genome of rice is one of the smallest genomes in monocots (2). Species with small genomes usually have small chromosomes. Rice chromosomes are indeed small and also difficult to prepare. Karyotype analysis in rice was almost impossible until the recent advent of modern chromosome preparation techniques. Consequently, the progress in rice cytogenetic studies was slower during the last century compared to other crops that have large chromosomes, such as wheat, barley, or maize. However, the small size of chromosomes and their moderate number provide advantages for making pachytene chromosome preparations in rice. The introduction of meiotic pachytene chromosomes into cytological studies of rice has helped overcome most of the difficulties caused by the small size of chromosomes.

Traditionally, two major methods have been used for staining rice pachytene chromosomes. One uses acetocarmine (3), while the other uses Giemsa (4). Both of these methods can generate good results and differentiate well between heterochromatins and euchromatins, which is very helpful for chromosome identification.

Wojciech P. Pawlowski et al. (eds.), *Plant Meiosis: Methods and Protocols*, Methods in Molecular Biology, vol. 990,
DOI 10.1007/978-1-62703-333-6_13, © Springer Science+Business Media New York 2013

Fluorescence in situ hybridization (FISH) is a newly developed technique for rice cytogenetic studies. Many tandem repetitive sequences have been identified in the rice genome, including 45S rDNA, 5S rDNA, CentO (a centromere-specific repeat), and Os48 (a subtelomeric repeat) (5, 6). These repetitive sequences can generate very specific FISH signals for characterizing rice chromosomes. In addition, two sets of rice chromosome-arm-specific cytogenetic markers based on rice bacterial artificial chromosomes (BACs) have been developed, which are very useful for identifying and tracing specific chromosome segments (7, 8).

Immunostaining is an in situ technique to identify a cellular constituent by detecting specific antibody-antigen interaction (9). Central to this technique is the use of an antibody tagged with a visible label, to link a cellular antigen to a stain that can be readily visualized under a microscope. As most meiotic proteins appear and disappear at specific times on meiotic chromosomes, detecting their presence at the right stage of meiosis is a good way to demonstrate their biological involvement.

Since the completion of the rice genome sequence, functional genomic studies have very popular. So far, hundreds of rice genes related to agronomic traits, development, and metabolism have been identified and their biological functions have been characterized. There are also several advantages for using rice for the investigation of meiosis, for example, the ease of collecting meiocyte samples, generating meiotic mutants, and conducting genetic transformation. These advantages make rice the third model plant species for the study of molecular mechanisms of meiosis, following maize and *Arabidopsis*. So far, almost a dozen genes related to meiosis have been identified in rice (10–14). In this chapter, I present protocols for basic meiotic chromosome preparation, FISH analysis, as well as immunocytology in rice that have been developed and successfully used in the past few years.

2 Materials

2.1 Chromosome Spreads for Basic Cytology

1. Carnoy's fixative: 3 volumes of 100% ethanol, 1 volume of glacial acetic acid.

2. 1% Acetocarmine solution: add 1 g carmine to 100 ml of 45% acetic acid in a 500 ml boiling flask and boil for 8 h on a hot plate with an attached reflux column. After boiling, place on ice immediately and cool rapidly. Store at room temperature until use.

3. Counterstaining solution: dissolve 10 mg of 4′, 6-diamidino-phenylindole (DAPI) in 10 ml of deionized water to make a stock solution. Dispense into aliquots and store at –20°C. Add 10 μl of the DAPI stock solution to 1 ml of Vectashield antifade solution (Vector Laboratories, Burlingame, CA).

**2.2   Fluorescent
In Situ Hybridization**

2.2.1   Probe Preparation

1. 10× Nick-translation buffer: 0.5 M Tris–HCl pH 7.5, 50 mM $MgCl_2$.

2. dNTPs (A,C,G): prepare a solution containing 0.5 mM dATP, 0.5 mM dGTP, and 0.5 mM dCTP.

3. 0.5 mM digoxigenin-dUTP or biotin-dUTP (see Note 1).

4. Sheared salmon sperm DNA: dissolve 10 mg of sheared salmon sperm DNA in 1 ml of deionized water.

5. 50% Dextran sulfate: dissolve 500 mg of dextran sulfate in 1 ml of deionized water.

*2.2.2   Hybridization,
Washing, and
Immunodetection*

1. 20× SSC: 3.0 M NaCl, 0.3 M sodium citrate.
   To make 1 L of 20× SSC, dissolve 175.3 g of NaCl and 88.2 g of sodium citrate $(2H_2O)$ in 900 ml of deionized water. Use 10 N HCl to adjust the pH to 7.0. Bring the volume to 1 L with deionized water. Store at room temperature after sterilizing by autoclaving.

2. 10× PBS: 1.37 M NaCl, 27 mM KCl, 100 mM Na_2HPO_4, 20 mM KH_2PO_4, pH 7.4.
   To make 1 L of 10× PBS, dissolve 80 g of NaCl, 2 g of KCl, 14.2 g of Na_2HPO_4, and 2.7 g of KH_2PO_4 in 900 ml of deionized water. Use 5 N HCl to adjust the pH to 7.4. Bring volume up to 1 L with deionized water. Store at room temperature after sterilizing by autoclaving.

3. 5× TNB: 0.5 M Tris–HCl pH7.5, 0.75 M NaCl, 2.5% blocking reagent (Boehringer Ingelheim, Ingelheim am Rhein, Germany).
   To make 10 ml of 5× TNB, mix 5 ml of 1 M Tris–HCl (pH 7.4), 0.4383 g of NaCl, and 0.25 g of the blocking reagent. Bring the volume up to 10 ml with deionized water and store at –20°C until use.

4. Denaturing solution: mix 7 ml of deionized formamide, 1 ml of 20× SSC, and 2 ml of deionized water and store at 4°C until use.

5. Stop solution: 0.2 M EDTA.

6. Detection solution with antibody: for each slide, mix 80 μl deionized water with 20 μl of 5× TNB buffer and then add 1 μl of the antibody. Prepare fresh just before use.

2.3   Immunostaining

4% Paraformaldehyde: dissolve 4 g paraformaldehyde in 95 ml of deionized water in a beaker with a stir bar. Cover the beaker and heat to 70°C on a hot plate. Add a few drops of 1 N NaOH until the solution clears. Cool on ice. Use 1 N HCl to adjust the pH to 7.0. Bring volume up to 100 ml with deionized water and store at 4°C until use.

3 Methods

3.1 Preparing Meiotic Chromosomes for Basic Cytology and FISH

1. Collect young panicles of rice at meiosis stage (see Note 2) and fix them with Carnoy's fixative solution overnight at room temperature. The tissue can be stored in the fixative at –20°C for a couple of months.

2. From the fixed panicle, take out a spikelet containing microsporocytes at the appropriate meiosis stage(s) (see Note 3) and put it on a slide. Dissect all six anthers from the spikelet using a pair of fine forceps in one hand and a needle in the other hand. Cut the anthers in half with the needle and add one small drop of acetocarmine on top of the anthers (see Note 4).

3. Squash the anthers with the needle quickly and completely (see Note 5), and cover them with a 22×22 mm coverslip.

4. Gently and evenly heat the slide containing the anthers over a flame of an alcohol lamp to near-boiling. At that time, both cytoplasm and chromosomes will be strongly stained, and the cells will be resistant to swelling.

5. Put one drop of 45% acetic acid on the left side of the coverslip, then absorb it out from the right side with a piece of filter paper. Repeat this step until the red color of the solution under the coverslip goes away.

6. Heat the slide gently again, and put it on a filter paper with the coverslip down. Finally, press the slide firmly with two fingers.

7. The slide is ready to be examined under a phase contrast microscope (Fig. 1a).

8. To proceed with the FISH protocol, soak the slide with coverslip on in liquid nitrogen and then remove the coverslip with a

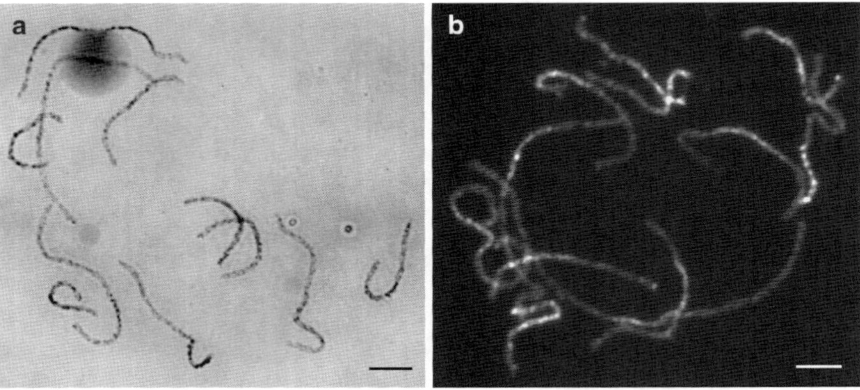

Fig. 1 Pachytene chromosomes of *indica* rice cv. Zhongxian 3037 stained with acetocarmine (**a**) and DAPI (**b**). Bars, 5 μm

razor blade. Dry the slide in an ethanol series (70, 90, and 100% ethanol, 5 min each wash) (see Note 6).

9. Add 10 µl of the Vectashield mount solution containing DAPI, and cover with a 24 × 32 mm coverslip. Put the slides (spread-side down) on a paper towel and allow the viscous DAPI solution to spread, minimizing air bubbles.

10. Visualize chromosomes under a fluorescent microscope. As rice chromosomes are very small, a microscope equipped with a 100× lens and a high-resolution/high-sensitivity CCD camera with appropriate filters should be used (Fig. 1b).

11. Freeze-dry slides and store in black plastic freeze boxes. The slides may be stored at –20°C for a couple of weeks.

3.2 Fluorescent In Situ Hybridization

3.2.1 Probe Labeling

1. Prepare the following reaction:

10× nick-translation buffer	5 µl
0.5 mM dNTP (A,C,G) solution	5 µl
0.5 mM Digoxigenin-dUTP or Biotin-dUTP	5 µl
Probe DNA	1 µg
DNA Polymerase I	12 U
DNase I	about 1 U (see Note 7)

2. Use deionized water to adjust the total volume to 50 µl. Mix carefully after each step of adding solutions and spin down at the end.

3. Incubate the reaction tube in a water bath at 15°C for 2 h.

4. Stop the reaction by adding 5 µl of stop buffer (0.2 M EDTA).

3.2.2 Hybridization of the Probe to Chromosomes

1. Prepare the hybridization mixture as follows:

Sheared salmon sperm DNA (10 mg/mL)	2 µl
Probe DNA	2 µl
Deionized formamide	10 µl
20× SSC	2 µl
50% dextran sulfate	4 µl

2. Denature this mixture by placing on an 80°C heating block for 5 min and then immediately chill the mixture on ice. Spin down briefly.

3. Add 100 µl of 70% formamide in 2× SSC on the dried slides (from Subheading 3.1, step 8) and cover the slides with 24 × 32 mm coverslips. Place slides on a plate (metal or glass) in an 80°C oven for 2.5 min.

4. Remove the coverslips by swinging the slides, and dip the slides immediately into a cold ethanol series (70, 90, and 100%, 5 min each at –20°C). Air-dry the slides.

5. Prepare the wet chamber (e.g., a plastic box with lid). Put a few layers of paper towels on the bottom of the box and a plastic holder on top of the paper towels. Pour a thin layer of water onto the paper towels.

6. Apply 20 µl of hybridization mixture to each slide and cover with a 24×32 mm coverslip. Seal the coverslip with rubber cement and place the slides in the wet chamber.

7. Incubate the wet chamber at 37°C overnight, a minimum of 6 h.

8. Peel the rubber cement from the slide using forceps. Dip the slides in a staining jar containing 2× SSC, slowly shaking the staining jar until the coverslips fall out from the slides.

9. Wash the slides using the following steps:

2× SSC, 42°C	10 min
2× SSC, room temperature	5 min
1× PBS, room temperature	5 min

10. After the 1× PBS wash, drain (but do not dry) the slides with a paper towel. Apply 100 µl of the detection solution on each slide, and cover the slide with a 24×32 mm coverslip. Place the slides in the same wet chamber as used before and incubate it at 37°C for 30 min.

11. Remove the coverslips by tilting the slides or dipping them in a beaker with 1× PBS.

12. Wash the slides in 1× PBS at room temperature three times.

13. Drain the slides on a paper towel. Add 10 µl of the Vectashield solution with DAPI, and cover them with a 24×32 mm coverslip.

14. Put the slides (spread-side down) over a paper towel and allow the viscous DAPI solution to spread, minimizing air bubbles.

15. Check the in situ hybridization results using a fluorescence microscope. FITC-labeled probes will emit yellow-green light. Rhodamine-labeled probes will emit red light (Fig. 2).

3.3 Immunostaining

3.3.1 Chromosome Spreads

1. Harvest fresh young panicles at a desired meiosis stage and fix them in 4% (w/v) paraformaldehyde for 10–30 min at room temperature. Wash the panicles briefly three times in 1× PBS at room temperature.

2. Dissect and squash anthers at the proper stage on a slide with 1× PBS solution. Cover the anthers with a coverslip.

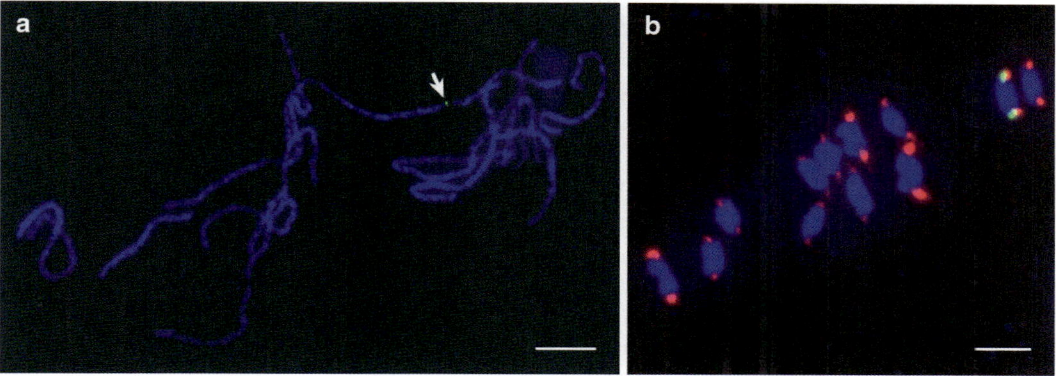

Fig. 2 FISH analysis of rice meiotic chromosomes. (**a**) Pachytene chromosome probed with G200, a 700 bp-long molecular marker located on chromosome 6. (**b**) Metaphase I chromosomes probed with the centromeric repeat CentO (*red*) and 5S rDNA (*green*). Chromosomes are stained with DAPI and are in *blue*. Bars, 5 μm

3. Put the slide on a filter paper with the coverslip side down. Press the slide firmly with two fingers.

4. Soak the slide in liquid nitrogen and remove the coverslip quickly with a blade. Dehydrate the slide through an ethanol series (70, 90, and 100%) prior to immunostaining.

3.3.2 Immunodetection

1. Dilute the desired primary antibody combination 1:500 in 1× TNB buffer.

2. Put 100 μl of the diluted antibodies onto the slide containing the squashed anthers and cover them with a 24×32 mm coverslip. Place the slides in the wet chamber and incubate at 37°C for 4 h.

3. Remove the coverslips by tilting the slides or dipping the slides in a beaker with 1× PBS.

4. Perform three rounds of washes in 1× PBS for 5 min each.

5. Dilute the desired combination of fluorophore-coupled secondary antibodies at 1:1,000 in 1× TNB buffer.

6. Put 100 μl of the diluted antibodies onto the slide and covered the slide with a 24×32 mm coverslip. Place the slides in the wet chamber and incubate at 37°C for 2 h.

7. Remove the coverslips as in step 3.

8. Perform three rounds of washes in 1× PBS for 5 min each.

9. Counterstain the chromosomes with DAPI in the antifade solution and cover the slide with a 24×32 mm coverslip.

10. Put the slides (spread-side down) on a paper towel to allow the viscous DAPI solution to spread. Minimize air bubbles.

11. Check the immunostaining results using a fluorescence microscope (Fig. 3).

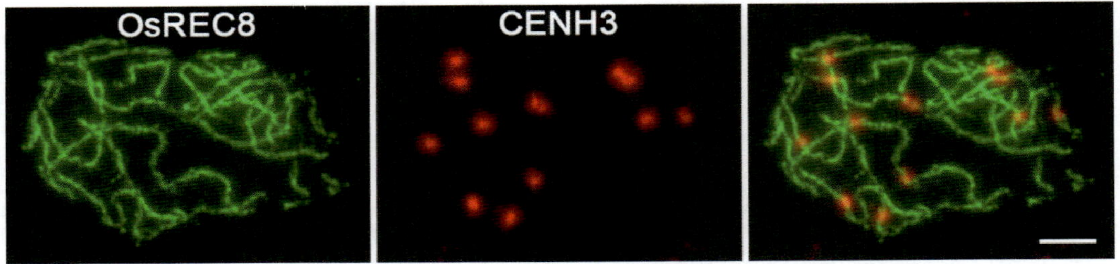

Fig. 3 Immunostaining of rice pachytene chromosomes with antibodies against the OsREC8 (*green*) and CENH3 (*red*) proteins. Bar, 5 μm

4   Notes

1. Instead of using straight 0.5 mM solutions of digoxigenin-dUTP or biotin-dUTP, a diluted solution made by mixing two volumes of 0.5 mM dTTP with one volume of 0.5 mM digoxigenin-dUTP or biotin-dUTP may be used—a 67% saving on the amount of labeled dUTP required for the experiments.

2. Use morphological characteristics to decide when to harvest rice panicles containing meiotic anthers. Normally, at the time of meiosis, the panicle is about 5–10 cm in length, and the span between the blade–sheath junctions of the flag leaf and the leaf below is around 0–2 cm.

3. Overall, spikelets in the upper part of a panicle are older than those in the lower part. However, on specific branches, the uppermost spikelet is the oldest one, while the next spikelet below it is the youngest one. For example, if there are five spikelets on a branch, 1 being the uppermost and 5 being the lowermost, their order from the youngest to the oldest will be 1, 5, 4, 3, and 2. For a branch with only three spikelets, the order will be 1, 3, 2 (Fig. 4).

4. If a slide is only to be used for phase contrast microscopy observations, a trace amount of $FeCl_3$ should be added to the Carnoy's fixative and a trace amount of $Fe(OH)_3$ should be added to the 1% acetocarmine solution. The Fe^{3+} ions facilitate differential staining of heterochromatins and euchromatins, particularly on pachytene chromosomes.

5. If this step is too long, over-staining may result.

6. If the slide is to be used for FISH analysis, steps 9–11 should be omitted. Instead, the slide may be immersed in 4% paraformaldehyde for 5 min for additional fixation, if necessary, before the tissue is dried in the ethanol series.

7. Obtaining the correct size of the final nick translation product is the most critical part of the labeling protocol. The best in

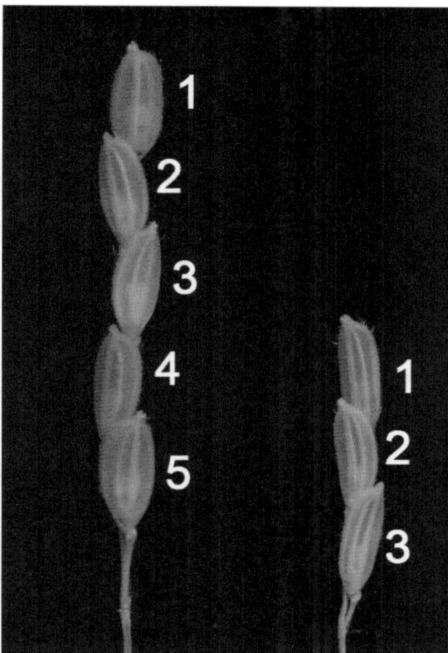

Fig. 4 Branches of a rice panicle. The numbers indicate the order of spikelets on the branch

situ hybridization results are obtained when the size of the labeled probe is around 200–600 bp. As DNase I may lose its activity during storage, the optimal dosage of DNase I may need to be determined experimentally every time to obtain the desired probe fragment size.

Acknowledgment

This work was supported by grants from the Ministry of Sciences and Technology of China (2011CB944602 and 2012AA100101).

References

1. Goff SA (1999) Rice as a model for cereal genomics. Curr Opin Plant Biol 2:86–89

2. Arumuganathan K, Earle ED (1991) Estimation of nuclear DNA content of plants by flow cytometry. Plant Mole Biol Rep 9:229–241

3. Wu HK (1967) Note on preparing of pachytene chromosomes by double mordant. Sci Agric 15:40–44

4. Kurata N, Omura T, Iwata N (1981) Studies on centromere, chromomere and nucleolus in pachytene nuclei of rice, Oryza sativa, microsporocytes. Cytologia 46:791–800

5. Cheng Z, Stupar RM, Gu M, Jiang J (2001) A tandemly repeated DNA sequence is associated with both knob-like heterochromatin and a highly decondensed structure in the meiotic pachytene chromosomes of rice. Chromosoma 110:24–31

6. Cheng Z, Dong F, Langdon T, Ouyang S, Buell CR, Gu M et al (2002) Functional rice

centromeres are marked by a satellite repeat and a centromere-specific retrotransposon. Plant Cell 14:1691

7. Cheng Z, Buell CR, Wing RA, Gu M, Jiang J (2001) Toward a cytological characterization of the rice genome. Genome Res 11:2133

8. Tang X, Bao W, Zhang W, Cheng Z (2007) Identification of chromosomes from multiple rice genomes using a universal molecular cytogenetic marker system. J Integr Plant Biol 49:953–960

9. Pawlowski WP, Golubovskaya IN, Timofejeva L, Meeley RB, Sheridan WF, Cande WZ (2004) Coordination of meiotic recombination, pairing, and synapsis by PHS1. Science 303: 89–92

10. Wang K, Tang D, Wang M, Lu J, Yu H, Liu J et al (2009) MER3 is required for normal meiotic crossover formation, but not for presynaptic alignment in rice. J Cell Sci 122:2055–2063

11. Wang M, Wang K, Tang D, Wei C, Li M, Shen Y et al (2010) The central element protein ZEP1 of the synaptonemal complex regulates the number of crossovers during meiosis in rice. Plant Cell 22:417–430

12. Che L, Tang D, Wang K, Wang M, Zhu K, Yu H et al (2011) OsAM1 is required for leptotene-zygotene transition in rice. Cell Res 21:654–665

13. Yu H, Wang M, Tang D, Wang K, Chen F, Gong Z et al (2010) OsSPO11-1 is essential for both homologous chromosome pairing and crossover formation in rice. Chromosoma 119:625–636

14. Nonomura K, Morohoshi A, Nakano M, Eiguchi M, Miyao A, Hirochika H et al (2007) A germ cell specific gene of the ARGONAUTE family is essential for the progression of premeiotic mitosis and meiosis during sporogenesis in rice. Plant Cell 19:2583–2594

Chapter 14

Analyzing Meiosis in Barley

James D. Higgins

Abstract

This chapter contains a detailed description of the immunological and cytological techniques developed to determine factors involved in crossover control during meiosis in barley. The immunological technique involves digesting fresh anthers followed by chromosome spreading to remove cytoplasm that causes unwanted background, whilst protecting the fragile chromosome structure from being compromised. Specific antibodies raised against meiotic proteins are then incubated with nuclei, detected with secondary antibodies conjugated to fluorescent dyes, and visualized with either wide-field or confocal microscopes. In the cytological technique, barley inflorescences are fixed, followed by dissecting out the anthers, digesting the cell walls and then spreading the meiotic chromosomes. Both techniques can be used in conjunction with specific DNA probes for fluorescent in situ hybridization (FISH) to label telomeres, centromeres, and ribosomal DNA; to identify DNA modifications such as 5-methylcytosine; and to detect the incorporation of DNA base analogues such as 5-bromo-2′-deoxyuridine (BrdU) or 5-ethynyl-2′-deoxyuridine (EdU) to be used for a meiotic time-course or assaying newly synthesized DNA. Although these techniques have been specifically developed for barley, they should be directly transferable to other cereal crop species such as wheat and rice.

Keywords Barley, Chromosomes, Cytology, BrdU, EdU, FISH, Immunolocalization, Meiosis

---

1   Introduction

Barley is one of the most suitable crop plants for cytogenetic studies being a diploid self-fertilizer with a low number $(2n = 2x = 14)$ of relatively large chromosomes (1). Previous advances in barley cytological techniques have involved chromosome banding (Giemsa C- and N-banding) (2, 3) and the use of short repetitive sequences such as GAA_7 probes to identify heterochromatic regions, used for karyotyping (4). Ribosomal DNA probes have also been used to characterize polymorphisms between varieties (5) as well as chromosome rearrangements or deletions (6). More recently bacterial artificial chromosomes (BACs) have been used to assist physical mapping and for developing a cytology-based recombination assay (7). However, the study of the early meiotic prophase stages

Wojciech P. Pawlowski et al. (eds.), *Plant Meiosis: Methods and Protocols*, Methods in Molecular Biology, vol. 990, DOI 10.1007/978-1-62703-333-6_14, © Springer Science+Business Media New York 2013

up to diakinesis has been considered impracticable due to technical limitations (1).

During the last decade our group has developed a range of molecular cytogenetical techniques for studying meiosis in the model plant *Arabidopsis*. These techniques have included cytological (8), immunological (9, 10) fluorescent in situ hybridization (FISH) (11, 12) and time-course assays (13). This chapter contains a detailed description of the techniques developed in *Arabidopsis* and adapted for barley in order to gain a better understanding of crossover control during meiosis. The main focus is to describe the immunological technique which has been optimized for barley. The aim of the immunological technique is to reveal the antibodies' epitope whilst removing excess protein that the antibody may non-specifically bind to thus causing background. It is essential to strike a balance between under/over cell wall digestion and under/over chromosome spreading. Too little digestion/spreading will reduce antibody penetration and leave excess protein (and background) whereas too much will destroy the large fragile barley chromosomes. Incubation times and reagents have been optimized in these protocols to achieve a high degree of antibody penetration whilst reducing unwanted background. Cytological techniques employed for barley are also described in detail. These are essentially the same as those used in *Arabidopsis*, except for differences that occur due to starting material. This chapter also provides protocols for analyzing DNA synthesis, DNA modifications, and probes used for FISH, both immunologically and cytologically for characterizing meiosis in barley.

2 Materials

2.1 Plants

Barley seeds are sown (cultivars Morex, Golden Promise, Optic, and Bowman have previously been used) in ≥6 cm diameter pots in soil-based compost and grown in dedicated growth chambers maintained at 22°C with a 16 h light cycle.

2.2 Immuno-localization

1. Meiocyte extraction medium: 2:1 mixture of citrate buffer and water. Prepare a 0.01 M citrate buffer (pH 4.5) solution by using 4.45 ml of 0.1 M sodium citrate and 5.55 ml of 0.1 M citric acid, made up to 100 ml with sterile deionized water. Dilute citrate buffer to meiocyte extraction medium working concentration with sterile deionized water, e.g., 200 μl of 0.01 M citrate buffer and 100 μl of sterile deionized water.

2. Digestion medium: dissolve 0.375 g sucrose and 0.25 g polyvinylpyrrolidone (MW 40,000; Sigma Aldrich, St. Louis, MO, USA) in 25 ml sterile deionized water. Aliquots of 1 ml should be dispensed and can be stored at –20°C.

3. Spreading medium: 1.5% Lipsol (SciLabware Ltd., Stone, Staffordshire, UK) in sterile deionized water.

4. Paraformaldehyde fixative: weigh out 4 g paraformaldehyde (EM grade) in the fume hood. Dissolve in 100 ml of sterile deionized water and add 4 drops of 1 M NaOH, pre-warmed to 60°C in the microwave. Stir the mixture on a magnetic stirrer for 1 h, or until it dissolves before filtering through Whatman paper. Adjust the pH to 8.0. The fixative can be stored for up to 1 week at 4°C.

5. Blocking buffer: dissolve 1% bovine serum albumen (BSA) with 0.1% Triton X-100 in 1× phosphate buffered saline (PBS). We tend to use prepared tablets, but PBS can also be made up from a 10× stock containing 1.37 M NaCl, 27 mM KCl, 100 m Na_2HPO_4, 18 mM KH_2PO_4, adjusted to pH 7.4). Autoclave before storage at room temperature.

6. Primary antibodies: make up to the relevant dilution, e.g., 1:100, 1:500, 1:1,000 in blocking buffer.

7. Washing solution: 1× PBS with 0.1% Triton X-100.

8. Secondary antibodies (e.g., anti-rat, anti-rabbit, anti-guinea pig, or anti-mouse antibodies, conjugated to FITC, Cy3, or Alexa Fluor (Invitrogen, Carlsbad, CA, USA) dyes): make up to the relevant dilution (generally 1:50 for FITC and 1:200 for Cy3) in 1% BSA and 1× PBS with 0.1% Triton X-100.

9. Counterstaining solution: 10 µl/ml of 4, 6-diaminido-2-phenlyinidole (DAPI) at 1 mg/ml in an antifade mounting medium, Vectashield (Vector Laboratories, Burlingame, CA, USA). Store DAPI as a stock solution at 1 mg/ml in sterile deionized water. Dispense in aliquots and store at –20°C.

2.3 Preparation of Slides for Basic Cytology, FISH Analysis, and Time-Course

1. Fixative: 3 parts of absolute ethanol: 1 part of glacial acetic acid. Prepare fresh fixative as required and discard at the end of the day.

2. 0.01 M Citrate Buffer: prepare a working solution of the buffer (pH 4.5) by using 4.45 ml of 0.1 M sodium citrate and 5.55 ml of 0.1 M citric acid, made up to 100 ml with sterile deionized water.

3. Stock digestion medium: dissolve 1% cellulase (Sigma Aldrich, St. Louis, MO, USA), 1% pectolyase (Sigma Aldrich, St. Louis, MO, USA) in a working solution of 0.01 M citrate buffer, pH 4.5. Store in aliquots at –20°C.

4. Digestion medium: mix 333 µl of the stock digestion medium with 667 µl of 0.01 M citrate buffer, pH 4.5.

5. 60% acetic acid: dilute glacial acetic acid with sterile deionized water.

6. Counterstaining solution: 10 µl/ml of 4, 6-diaminido-2-phenlyinidole (DAPI) at 1 mg/ml in an anti-fade mounting medium, Vectashield (Vector Laboratories, Burlingame, CA, USA). Store DAPI as a stock solution at 1 mg/ml in sterile deionized water. Dispense in aliquots and store at –20°C.

2.4 FISH Analysis

1. Probe labelling: use the nick translation labelling kit (Roche Diagnostics, Burgess Hill, UK) following the manufacturer's instructions. Use either biotin-16-dUTP or digoxigenin-11-dUTP (Roche Diagnostics, Burgess Hill, UK) as nucleotide conjugates for DNA labelling (see Subheading 2.5 for probes).

2. Hybridization mix: weigh out 1 g dextran sulfate (use MW 500,000), add 5 ml of deionized formamide, (see Note 1) and 1 ml of 20× SSC, make up to 7 ml with sterile deionized water. Dissolve at 65°C, cool and pH to 7.0. Aliquot into 1.5 ml microfuge tubes and store at –20°C.

3. Prepare 20 µl of probe mixture per slide: 14 µl of hybridization mix, 0.5–2 µl of labelled probe, and if necessary add sterile deionized water to 20 µl. Thus, the final hybridization mix consists of 50% deionized formamide, 2× SSC and 10% dextran sulfate pH 7.0.

4. Vulcanizing rubber solution (e.g., as found in bicycle wheel repair kits).

5. Make up post-hybridization washes: 3 Coplin jars of 50% formamide–2× SSC pH 7.0 (150 ml of deionized formamide, 30 ml of 20× SSC, and 120 ml of sterile deionized water), with 4 T buffer (4× SSC + 0.05% Tween 20) for all subsequent washes.

6. For detection of digoxigenin probes make up antibodies as either anti-digoxigenin-FITC or anti-digoxigenin rhodamine at 5 ng/µl in digoxigenin blocking solution shortly before use. The blocking solution is made with 4 T buffer and 0.5% digoxigenin blocking reagent (Roche Diagnostics, Burgess Hill, UK), centrifuged at $19,000 \times g$ for 5 min, and the supernatant stored in 1 ml aliquots at –20°C.

7. For biotin labelled probes use streptavidin-Cy3/FITC (Roche Diagnostics, Burgess Hill, UK) made up in biotin blocking solution. This solution is made with 4 T buffer and 5 g dried skimmed milk. Centrifuge at $19,000 \times g$ for 5 min and store the supernatant in 1 ml aliquots at –20°C.

8. Counterstaining solution: see Subheading 2.2, item 9.

2.5 DNA Probes for FISH

1. Telomere probe: use oligonucleotide sequences 5′ – tttagggtt-tagggtttagggtttagggtttagggg – 3′ and 5′ –ccctaaaccctaaac-cctaaaccctaaaccctaaa – 3′ (100 pmol each) in a 50 µl primary reaction as both primer and template. Perform PCR using

93°C, 30s; 55°C, 45 s; 72°C, 45 s; 30 cycles. Then conduct a secondary PCR using as template 1 µl from the primary PCR but replacing dTTP with dUTP-biotin/digoxigenin to directly label the probe.

2. Centromere probe: use oligonucleotide sequences 5′ –agggagagggagagggagagggagag– 3′ and 5′ –ctccctctccctctccctctccctct – 3′ (14) as in Subheading 2.5, item 1.

3. 5*S* ribosomal DNA: we use plasmid pCT4.2 containing the 5*S* rDNA gene from *Arabidopsis thaliana* as a 500 bp insert cloned in pBlu (ABRC).

4. 45*S* ribosomal DNA: we use clone pTa71 (15) containing a 9 kb *Eco*RI fragment of *Triticum aestivum* consisting of the 18*S*–25*S* rRNA genes and the spacer regions.

2.6 DNA Base Incorporation and DNA Modification Detection

1. Needle for BrdU/EdU injection (25 G×1 in. per 100 (ND400), Becton Dickinson, Oxford, UK) attached to a 1 ml plastic syringe.

2. 5-Bromo-2′-deoxy-uridine Labelling and Detection Kit I (Roche Diagnostics, Burgess Hill, UK) and/or the Click-iT EdU Alexa Fluor 594 HCS assay kit (Invitrogen, Carlsbad, CA, USA). EdU (Invitrogen, Carlsbad, CA, USA) can be ordered separately as larger volumes may be required. Make up EdU to 10 mM in sterile 1× PBS and store at –20°C.

3. An anti-5-methylcytosine antibody (Cat. No. MAb-31HMC-100, Diagenode, Liege, Belgium) raised in mouse was used to detect methylated DNA.

---

3 Methods

3.1 Plant Material

To undertake a meiotic analysis of barley, the correct plant material first has to be collected. In our growth conditions, it takes ~6 weeks from sowing the seeds to collecting inflorescences of the correct size. The plants are ~40 cm in height and the inflorescences are located within the stem, usually between the top two lateral leaves (see Note 2). The optimum size of inflorescences (1.5–2 cm) may be dissected out from the stems using a sharp pair of scissors.

3.2 Immuno-localization

1. Harvest inflorescences and place them on to moist 9 cm Ø filter paper in a 9 cm Ø Petri-dish to prevent drying out.

2. Using a dissecting microscope, calibrate the graticule so that 10 bars = 1 mm and, excise anthers (0.4–1 mm length) from the inflorescence with watchmakers forceps and a fine mounted needle.

3. Take a range of anther sizes and transfer ~10 anthers to 2 µl extraction buffer (see Note 3) on a cavity slide.

4. Cut the anthers transversely with a razor blade (or scalpel) and squeeze out the meiocytes with a thick (1 mm Ø) mounted needle.

5. Add 6 μl of water and 6 μl of digestion medium to the meiocytes, then mix with forceps and a thick (1 mm Ø) mounted needle.

6. Incubate slide at 37°C for 2 min in a closed moist container to prevent drying out of cells.

7. To a clean glass slide (see Notes 4 and 5) add 10 μl of 1.5% Lipsol (see Note 6).

8. Cut the end off a yellow tip and pipette 10 μl of the digested meiocytes onto the slide with the 10 μl of 1.5% Lipsol and gently spread with a fine mounted needle.

9. Add 20 μl of 4% paraformaldehyde to the cells and mix with a pipette tip, then allow to air dry in a fume hood.

10. When dry (~60 min), add 50 μl of blocking solution containing primary antibodies at a preferred concentration directly to the slide (see Subheading 2.1, items 5 and 6).

11. Cover slides with parafilm by placing on top and incubate at 37°C for 30 min or overnight at 4°C in a sealed plastic container with damp tissue paper to prevent drying out.

12. Wash slides in washing solution (2 × 5 min).

13. Drain off excess wash buffer by standing on tissue paper for ~2 min.

14. Add secondary antibodies and incubate at 37°C for 30 min.

15. Then wash as in step 12.

16. Drain off excess washing solution by standing on tissue paper for ~2 min and then add an appropriate mounting medium e.g., DAPI in Vectashield (Fig. 1a).

3.3 Cytological Preparation

1. Place harvested barley inflorescences (see Subheading 3.2, step 1) into glass vials containing 5–20 ml of fix solution.

2. Add fresh fix solution after 1 h then leave for at least 24 h.

3. Dissect anthers (0.5–1.5 mm length) from inflorescences with Watchmaker's forceps and a fine mounted needle.

4. Replace the fixative first with the citrate working buffer (2 × 5 min) and then with the working solution of enzymes and incubate in a humidified atmosphere (e.g., a sandwich box containing damp tissues) at 37°C for 30 min. Then, remove the enzyme solution and replace with cold (4°C) ~0.5 ml of sterile distilled water to stop the reaction.

5. Place a few anthers (~5) onto a slide, with a minimum of water and quickly macerate with a mounted needle, ensuring that the material does not dry out.

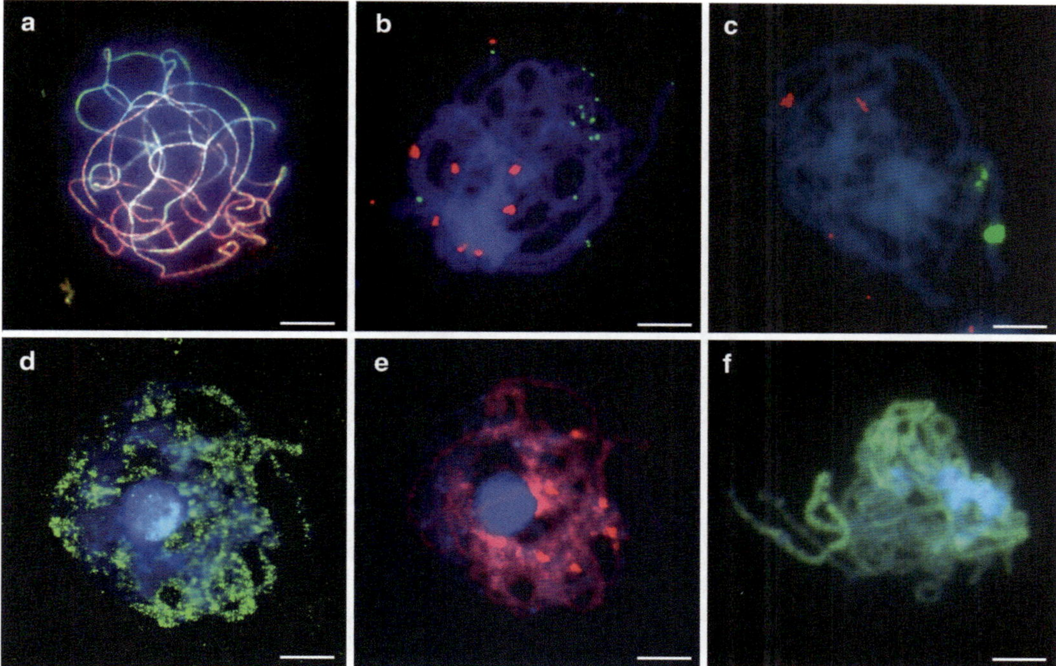

Fig. 1 Analyzing meiosis in barley. (**a**) A barley pachytene nucleus obtained using the immunocytological technique. The synaptonemal complex central protein, ZYP1 is labelled with CY3 (*red*) and the axis-associated protein, ASY1 is labelled with FITC (*green*). The chromosomes are counterstained with DAPI (*blue*). Scale bar, 10 μm. (**b**) A barley late-zygotene nucleus obtained using the cytological technique, in conjunction with FISH. The centromeres are labelled with Cy3 (*red*) and the telomeres labelled with FITC (*green*) as described in the methods. The chromosomes are counterstained with DAPI (*blue*). Scale bar, 10 μm. (**c**) A barley late-zygotene nucleus obtained using the cytological technique in conjunction with FISH. Ribosomal 5*S* and 45*S* DNA probes are labelled with Cy3, (*red*) and FITC (*green*) respectively, as described in the methods. The chromosomes are counterstained with DAPI (*blue*). Scale bar, 10 μm. (**d**) A barley late-zygotene nucleus obtained using the cytological technique showing BrdU labelled with FITC (*green*) that has been incorporated into the chromosomal DNA during meiotic S-phase and detected after 36 h. The chromosomes are counterstained with DAPI (*blue*). Scale bar, 10 μm. (**e**) A barley late-zygotene nucleus obtained using the cytological technique showing EdU labelled with Alexa Fluor 594 (*red*) incorporated into the chromosomal DNA during meiotic *S*-phase and detected after 36 h. The chromosomes are counterstained with DAPI (*blue*). Scale bar, 10 μm. (**f**) A barley late-zygotene nucleus obtained using the cytological technique showing detection of 5-methylcytosine labelled with FITC (*green*). The chromosomes are counterstained with DAPI (*blue*). Scale bar, 10 μm

6. Add 7 μl of 60% acetic acid, mix with the material on the slide and place on a hot block at 45°C for up to 30s.

7. Place the slide on the bench and add 2 × 200 μl fixative as a ring around the material. The preparation is then dried with a commercial hair dryer.

8. The slides are now ready for basic cytology after mounting in 7 μl DAPI in Vectashield.

3.4 Fluorescence In Situ Hybridization

1. Prepare required probes (see Subheading 2.5).

2. To use FISH in conjunction with immunolocalization it is first necessary to incubate a primary antibody (to a slide prepared

described in Subheading 3.2), and then wash (2× 5 min with 1× PBS) before adding a secondary antibody conjugated to biotin.

3. After incubating with the secondary antibody (30 min at 37°C or overnight at 4°C) and washing (2× 5 min in 1× PBS), the slides are ready for hybridization of the probes.

4. Cytological slides prepared in Subheading 3.3 can be used directly for denaturation and hybridization of probes.

5. Denaturation of the probe and chromosomes: place 20 μl of the probe mixture on to a slide with a coverslip (22×22 mm) and seal with vulcanizing rubber solution. The slides are then heated on a hotplate at 75°C for 4 min.

6. Hybridization: Incubate the slides in a sealed plastic container with damp tissue paper at 37°C overnight.

7. Post-hybridization washes: remove the rubber solution and coverslips using fine forceps. Wash the slides 3×5 min in 50% formamide–2× SSC at 45°C, then once in 2× SSC at 45°C for 5 min, then once in 4 T buffer at room temperature for 5 min.

8. Labelling detection: Add either anti-dixoxigenin- or streptavidin-fluorescent antibodies to the slides (50 μl per slide), cover with parafilm and incubate in a sealed plastic container with damp tissue paper at 37°C in the dark for 30 min. Remove the parafilm and wash the preparations in the washing solution 3×5 min.

9. Counterstain slides with 7 μl DAPI in Vectashield staining solution.

10. The FISH preparations can be viewed with a fluorescence microscope having filters for DAPI, Texas Red, Cy3 and FITC and equipped with an image capture and analysis system (see Fig. 1b, c).

3.5 DNA Base Incorporation and DNA Modification Detection

The incorporation of the DNA base analogue 5-bromo-2-deoxyuridine (BrdU) into newly synthesized DNA has been developed in *Arabidopsis* by Armstrong et al. (13) and that methodology is described in detail in Chapter 12. The detection steps in barley are similar to *Arabidopsis* but introducing the DNA base analogues into replicating cells is different (see Note 7).

1. Incorporation of (10 mM) BrdU and/or (10 mM) 5-ethynyl-2′-deoxyuridine (EdU) into meiotic S-phase cells is most successful when injecting with a fine needle and syringe directly into the inflorescence, or just above (see Notes 8–10).

2. Material can be collected at an appropriate time post-injection and inflorescences either fixed for cytological analysis or used fresh for immunolocalization.

3. If used for immunolocalization, follow the slide making protocol (Subheading 3.2) and use a secondary antibody conjugated to biotin (as in Subheading 3.4, step 2).

4. Add 20 µl of hybridization buffer (from Subheading 2.4, item 3) to each slide and place on a hot block at 70°C for 2 min (as in Subheading 3.4, step 4).

5. Directly transfer to a Coplin jar with prechilled (to 4°C) 1× PBS and leave for 5 min to snap chill the chromosomes (see Note 11).

6. Then remove cover-slips and wash for 2×5 min in 1× PBS at room temp.

7. For BrdU, follow manufacturers' instructions for detection.

8. If using EdU together with BrdU, then detect EdU afterwards according to the manufacturers' instructions (see Fig. 1d, e).

9. For detecting methylated DNA follow the procedure for immunolocalization as in Subheading 3.5, steps 3 and 4, and for cytological detection follow Subheading 3.5, steps 4–6. Incubate slides with the appropriate 5-methylcytosine primary antibody (30 min at 37°C) followed by washes (2×5 min in 1× PBS) and then detect with a secondary antibody fluorescent conjugated to a fluorescent dye (30 min at 37°C) followed by washes (2×5 min in 1× PBS) (see Fig. 1f).

4   Notes

1. We deionize formamide by mixing 200 ml formamide with 10 g mixed resin beads (PlusOne Amberlite IRN-150 L, GE Healthcare, Piscataway, NJ, USA) on a stirrer for at least 1 h. Filter and store at 4°C.

2. There is a large degree of variability between plant height and size of inflorescence. Therefore, it is advised either to use a noninvasive method such as pressing the inflorescence with thumb and forefinger to gain an idea of size, or to remove some of the stem material by forceps and scissors.

3. The correct extraction buffer is required to neutralize the extracelluar matrix components in the anther. It appears that a slightly acidic buffer is most effective.

4. It is best to pre-clean glass slides to reduce material from being washed off. Slides can be cleaned simply by sequential washing in 100% acetone (10 min), sterile distilled water (5 min), and 100% ethanol (2 min). The slides can then be left to dry at room temperature or dried with a hair-dryer.

5. For immunolocalization, I have a preference for the Superfrost Plus glass slides (Fisher Scientific, Suwanee, GA, USA) although ordinary frosted slides work successfully.

6. Optimum Lipsol concentrations lie between 1 and 2%. Early stages, such as G2 and leptotene, are more suited to 1.5–2%

Lipsol, whereas later stages, such as zygotene and pachytene, are better with 1–1.5% Lipsol.

7. Cutting stems and placing into BrdU was not successful in barley due to the inflorescence rapidly increasing in size in vivo but not in vitro.

8. It is sensible to wear goggles and gloves during this step as the BrdU/EdU is capable of coming out at high pressure.

9. The volume of BrdU/EdU injected into the plant is variable (usually 200–1,000 µl). Injection is stopped when drops of liquid appear at the top of the plant.

10. There is a rapid uptake of BrdU/EdU into replicating cells and incorporation can be detected 30 min post-injection.

11. This is necessary to prevent the chromosomal DNA from re-annealing, thus allowing the anti-BrdU antibody to bind to BrdU incorporated into the chromosomal DNA.

References

1. Taketa S, Linde-Laursen I, Künzel G (2003) Cytogenetic diversity. In: von Bothmer R, van Hintum T, Knüpffer H, Sato K (eds) Diversity in Barley (Hordeum vulgare). Elsevier, Amstedam, pp 97–119

2. Linde-Laursen I (1975) Giemsa C-banding of chromosomes of Emir barley. Hereditas 81:285–289

3. Islam AKMR (1980) Identification of wheat-barley addition lines with N-banding of chromosomes. Chromosoma 76:365–373

4. Pedersen C, Rasmussen SK, Linde-Laursen I (1996) Genome and chromosome identification in cultivated barley and related species of the Triticeae (*Poaceae*) by in situ hybridization with the GAA-satellite sequence. Genome 39:93–104

5. Taketa S, Harrison GE, Heslop-Harrison JS (1999) Comparative physical mapping of the 5S and 18S-25S rDNA in nine wild Hordeum species and cytotypes. Theor Appl Genet 98:1–9

6. Gecheff K, Hvarleva T, Georgiev S, Wilkes T, Karp A (1994) Cytological and molecular evidence of deletion of ribosomal-RNA genes in chromosome-6 of barley (*Hordeum vulgare*). Genome 37:419–425

7. Phillips D, Nibau C, Ramsay L, Waugh R, Jenkins G (2010) Development of a molecular cytogenetic recombination assay for barley. Cytogenet Genome Res 129:154–161

8. Ross KJ, Fransz P, Jones GH (1996) A light microscopic atlas of meiosis in *Arabidopsis thaliana*. Chromosome Res 4:507–516

9. Higgins JD, Armstrong SJ, Franklin FCH, Jones GH (2004) The *Arabidopsis* MutS homolog AtMSH4 functions at an early step in recombination: evidence for two classes of recombination in *Arabidopsis*. Genes Develop 18:2557–2570

10. Higgins JD, Sanchez-Moran E, Armstrong SJ, Jones GH, Franklin FCH (2005) The *Arabidopsis* synaptonemal complex protein ZYP1 is required for chromosome synapsis and normal fidelity of crossing over. Genes Develop 19:2488–2500

11. Sanchez-Moran E, Armstrong SJ, Santos JL, Franklin FC, Jones GH (2001) Chiasma formation in Arabidopsis thaliana accession Wassileskija and in two meiotic mutants. Chromosome Res 9:121–128

12. Armstrong SJ, Franklin FCH, Jones GH (2001) Nucleolus-associated telomere clustering and pairing precede meiotic chromosome synapsis in *Arabidopsis thaliana*. J Cell Sci 114:4207–4217

13. Armstrong SJ, Franklin FCH, Jones GH (2003) A meiotic time course for *Arabidopsis thaliana*. Sex Plant Reprod 16:141–149

14. Nasuda S, Hudakova S, Schubert I, Houben A, Endo TR (2005) Stable barley chromosomes without centromeric repeats. Proc Natl Acad Sci USA 102:9842–9847

15. Gerlach WL, Bedbrook JR (1979) Cloning and characterization of ribosomal RNA genes from wheat and barley. Nucleic Acids Res 7:1869–1885

Part II

Cytological Techniques for Electron Microscopy

Chapter 15

Preparing SC Spreads with RNs for EM Analysis

Lorinda K. Anderson and Stephen M. Stack

Abstract

Recombination nodules (RNs) are associated with synaptonemal complexes (SCs) during early prophase I of meiosis. RNs are too small to be resolved by light microscopy and can be observed directly only by electron microscopy. The patterns of RNs on SCs can be analyzed using three-dimensional reconstructions of nuclei using serial thin sections, but this method is time consuming and technically difficult. In contrast, spreads of SCs are in one plane so all RNs in each set can be visualized simultaneously, and the patterns of both early and late nodules (ENs and LNs) can be analyzed far more easily than using sections. Here, we describe methods for preparing spreads of SCs and RNs from tomato primary microsporocytes on plastic-coated slides for visualization by transmission electron microscopy (TEM).

Keywords Electron microscopy, Tomato, Synaptonemal complex, Recombination nodules, *Solanum lycopersicum*

1 Introduction

Synaptonemal complexes (SCs) and recombination nodules (RNs) were discovered using transmission electron microscopy (TEM) of thin-sectioned nuclei in prophase I of meiosis (1–3). Analysis of synapsis and RNs on SCs at that time required technically demanding and tedious three-dimensional reconstructions of serially sectioned nuclei (3–5). Fortunately, easier methods for preparing SCs and RNs for analysis were soon developed for animals (6, 7) and plants (8–10). These methods involve spreading chromosomes (SCs) from living cells using either surface tension or hypotonic swelling. The expansion of chromatin separates the chromosomes and permits visualization of SCs and RNs in two dimensions by TEM (Figs. 1 and 2).

Here, we describe a technique for spreading SCs from the model plant species *Solanum lycopersicum* (tomato) and the use of SC spreads with RNs for investigation of synapsis and recombination

Wojciech P. Pawlowski et al. (eds.), *Plant Meiosis: Methods and Protocols*, Methods in Molecular Biology, vol. 990, DOI 10.1007/978-1-62703-333-6_15, © Springer Science+Business Media New York 2013

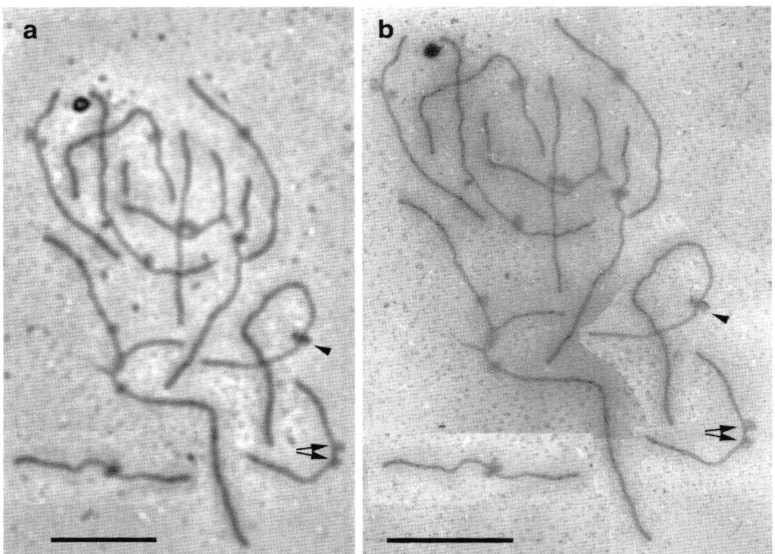

Fig. 1 SC spreads visualized by light and electron microscopy. The same SC spread from an F1 hybrid of two wild tomato species (*Solanum penellii* and *S. habrochaites*) visualized by (**a**) light microscopy using phase contrast and (**b**) EM. One kinetochore is indicated with an *arrowhead*, and some chromosomes have mismatched kinetochores (*double arrows*). *Bright flecks* and *lines* that are visible by phase-contrast microscopy are often less noticeable by EM. One SC of the set was moved to the *lower left* to make each figure more compact. Each bar equals 10 μm

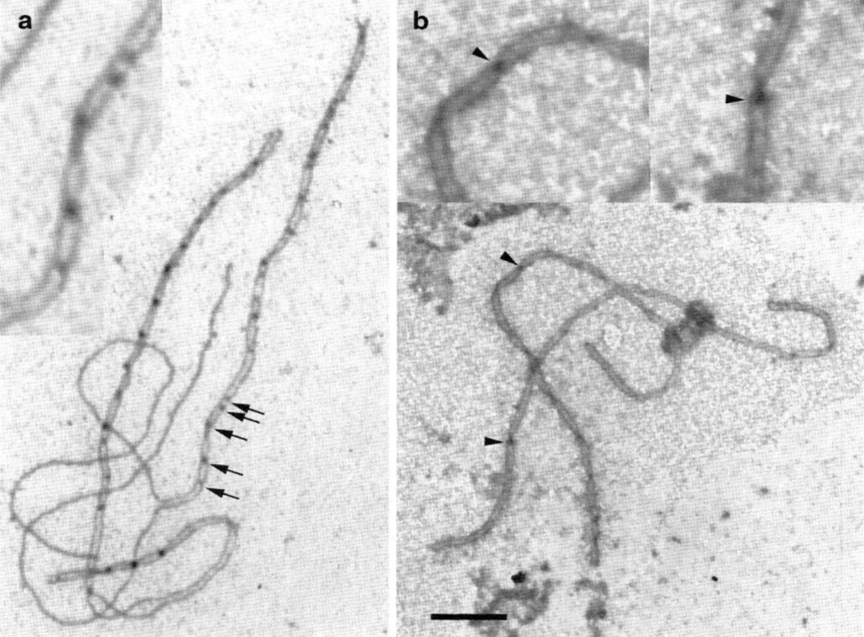

Fig. 2 TEM of tomato SCs. SC spreads from cultivated tomato in zygotene (**a**) and pachytene (**b**). Many ENs (some indicated with *arrows* and shown at 2.5 times higher magnification in the inset at *upper left*) are present on SC segments at zygotene. At pachytene, each of these two SCs has only one RN (*arrowheads* and shown at 2.5 times higher magnification in insets above). ENs and LNs are easier to identify in negatives by eye using an ×8 loupe magnifier than in printed images. The zygotene SC in (**a**) was stained with uranyl acetate and lead citrate (12). Bar equals 10 μm

by electron microscopy. With minor variations, this procedure can be applied to other plant species, and we have successfully used it to prepare SC spreads and analyze RNs from maize (*Zea mays*), lily (*Lilium longiflorum*), onion (*Allium cepa*), plateau spiderwort (*Tradescantia edwardsiana*), and the lower vascular plant, whisk fern (*Psilotum nudum*) (11–13). These chromosome spreads are also suitable for immunolocalization and fluorescence in situ hybridization (14–16).

---

2   Materials

2.1   *Microscope Slide Preparation*

1. 25 mm by 75 mm glass microscope slides frosted at one end on one side.
2. Kimwipes or other lint-free paper towels.
3. 0.3 g of Falcon plastic (broken from new petri dishes).
4. 50 ml of water-free dichloroethane.
5. 100 ml bottle with cap.
6. Ultrasonic cleaner.
7. 100 ml glass graduated cylinder.
8. Binder clip and lint-free string.
9. Plastic slide box (for 25 slides). Most of the bottom of the box needs to be removed so that only a narrow rim on each side of the base remains to keep slides from falling through.
10. Glow discharge apparatus large enough to hold the slide box. A glow discharge apparatus can be a commercially available unit (e.g., Electron Microscopy Sciences, Hatfield, PA, USA or Ladd Research, Williston, VT, USA), but it will need to be modified to deliver alternating current. Alternatively, a similar apparatus can be assembled from parts (17).

2.2   *Preparing Protoplasts*

1. Healthy, blooming tomato plants (see Note 1).
2. Dissecting microscope with an ocular micrometer.
3. Tools: medium fine-tipped forceps, a scalpel with a small blade (#11, Feather brand, Ted Pella, Redding, CA, USA), a sharp chisel-shaped dissecting needle, and a sharp pointed dissecting needle.
4. Digestion medium: potato culture medium powder (Carolina Biological Supply Company, Burlington, NC, USA) dissolved in distilled water as directed by the supplier. Use 62.5 ml of this solution and dilute to 100 ml with distilled water

(see Note 2). Divide the diluted solution into 1 ml aliquots, freeze and store at –20°C. To prepare digestion medium, mix 1 ml of this diluted potato culture medium with 1 ml of 2.5 mM $CaCl_2$, 1 ml of 1% aqueous potassium dextran sulfate (freshly prepared, 0.01 gm in 1 ml distilled water), and 2 ml of distilled water. Add 0.64 g of mannitol (Sigma Aldrich, St. Louis, MO, USA) and 0.1 g of polyvinylpyrrolidone (PVP-10, Sigma Aldrich, St. Louis, MO, USA) to the liquid mixture and adjust pH to 5.1 using 0.1 N KOH (and 0.1 N HCl, if needed). Avoid the use of sodium ions that are deleterious to the cells.

5. Desalted, lyophilized cytohelicase (Sigma Aldrich, St. Louis, MO, USA). Desalting greatly improves the success and reproducibility of spreading, probably by removing any sodium salts. Desalt the enzyme using a Sephadex G25 column, but do not use a UV-monitored fraction collector because the enzyme is light-sensitive. After freeze-drying, store at –20°C in amber microcentrifuge tubes.

6. 25 mm by 75 mm glass depression slide.

7. Incubation dish to maintain slides in a humid environment over water. We use plastic petri dishes with a bent glass bar in the bottom.

8. A scalpel with a #11 blade.

2.3 Spreading Synaptonemal Complexes

1. Bursting medium: aliquot 500 µl volumes of aqueous 0.05% IGEPAL CA-630 (Sigma Aldrich, St. Louis, MO, USA) and store at –20°C. Just before use, add 5 µl of freshly prepared aqueous 0.1% potassium dextran sulfate and 30 µl of 4% formaldehyde (see Subheading 2.3, item 2).

2. 4% formaldehyde (PFA): add 4 g of paraformaldehyde powder and 1 ml of 1 N NaOH to 95 ml of deionized or distilled water in a beaker and heat with stirring until the solution clears (keep solution temperature below 60°C). Cool to room temperature, add 1 ml of 0.05 M sodium borate and adjust pH to 8.5–8.7 with 1 N HCl. Aliquot into 15–20 ml volumes and store at –20°C. If precipitate remains after thawing, warm the solution slightly until the solution clears.

3. Glass micropipettes with aspirator tube assembly (Fisher Scientific, Suwanee, GA, USA). Use a flame to draw capillary tubes (size 1.5, 1 mm by 100 mm) to make micropipettes with tips ~0.13 mm in outside diameter. The tip edges should be smooth to avoid damage to cells. Smoothing can be done using a fine sharpening stone. Before each experiment, siliconize the interior of a micropipette by drawing up and expelling 5% methylchlorosilane in chloroform then rinsing the pipette with distilled water.

4. Medical nebulizer apparatus designed for inhalation therapy and an air compressor for the nebulizer (e.g., DeVilbiss Healthcare, Somerset, PA, USA).

5. Colored fingernail polish.

6. Photo-Flo 200 (Kodak, Rochester, NY, USA) diluted with deionized or distilled water to 0.4%.

2.4 Post-stain for Electron Microscopy

1. DNase I digestion buffer (10× stock): 100 mM Tris–HCl, 25 mM $MgCl_2$, 50 mM $CaCl_2$, pH 7.5. Dilute 1:10 with water just before use.

2. DNase I (Sigma Aldrich, St. Louis, MO, USA): 1 mg/ml in 50% glycerol. Store at –20°C.

3. Formaldehyde/Glutaraldehyde fixative (FG): mix 10 ml of 8% glutaraldehyde (Sigma Aldrich, St. Louis, MO, USA), 20 ml of 4% PFA (see Subheading 2.3, item 2), and 10 ml of 0.12 M sodium phosphate buffer (pH 7.0). Adjust the pH to 7.5–8.0 with 1 N NaOH.

4. Phosphotungstic acid (PTA) stain:

    (a) Dissolve 1 g of PTA in 25 ml of distilled water. This stock keeps for several days in a closed bottle at 4° in the dark.

    (b) Prepare the amount of staining solution needed by mixing one part aqueous PTA with 3 parts 95% EtOH.

2.5 Transferring Plastic to Grids

1. Center-marked 50- or 75-mesh "finder" grids (1GC50 or 1GC75; Ted Pella, Redding, CA, USA).

2. Grid glue: put about 20 ml of dichloroethane in a small glass bottle and swish a clean 2–3 cm long piece of cellophane tape in the solvent for about 1 min. Remove tape and excess blobs of adhesive. Store grid glue at room temperature for several weeks.

3. Eyelash tool: sharpen one end of a wooden applicator stick and glue a clean eyelash to the sharp end.

4. Hydrofluoric acid (HF): 1% in water. HF acid dissolves glass, so it must be kept in a plastic bottle. HF acid can be absorbed through the skin, so avoid skin contact.

5. Aqueous 5% acetic acid.

6. Parafilm.

7. Bowl of distilled water.

3 Methods

3.1 Preparing Plastic-Coated Microscope Slides for SC Spreading

1. Dissolve plastic: place 0.3 g of Falcon plastic (see Note 3) in a glass bottle containing 50 ml of dichloroethane, close the bottle with a screw cap, and sonicate to dissolve plastic. Any water contamination of the solvent will cause holes in the plastic film,

so use phase-contrast microscopy to check the first slide for the presence of holes. Make this solution fresh each time.

2. Wipe slide: Hold the slide by the frosted end onto a table with the fingers of one hand, and repeatedly wipe the slide from the frosted end to the other end with a small pad of dry Kimwipes until the glass seems relatively slick.

3. Apply plastic to slide: Select an area free from air drafts to avoid uneven drying of the plastic. Pour the plastic solution into a clean 100 ml glass graduated cylinder, tie a string to a small binder clip, and attach the clip to the frosted end of a freshly wiped slide. Remove any particles from the slide that could contaminate the plastic solution. Lower the slide into the graduated cylinder until it is submerged just above the frosting, but do not let the liquid touch the clip. Pull the slide out of the liquid quickly, and then slowly (over ~10 s) draw the slide straight up and out of the graduated cylinder. Hold the slide vertically with its bottom touching the top edge of the cylinder until the slide is completely dry. Remove the clip and place slide in a rack. Be sure not to touch the plastic surface at any time (see Note 4).

4. Testing plastic thickness: To have good resolution by EM, it is essential that the plastic not be too thick. To test the thickness, use a sharp dissecting needle to scratch the plastic film to make a ~2 cm circle. Slowly push the slide at a low angle, scratched surface up, into a bowl filled with water. The water will displace the hydrophobic plastic circle from the hydrophilic glass surface. After the circle of plastic is floating on the water, the interference color of the plastic indicates its thickness. Plastic with the best thickness is a silver-grey. Plastic with a gold color is too thick for good EM resolution, and plastic with a grey color is so thin that it easily breaks. The thickness can be modified by adjusting the length of time the slide drains in the graduated cylinder (more time = thinner plastic). If necessary, change the concentration of the plastic solution to obtain the correct plastic thickness (see Note 5).

5. Storing plastic-coated slides: After air-drying, slides can be stored indefinitely in a wooden slide box. Storage in some plastic slide boxes causes the plastic film to be difficult to lift from the glass slide.

3.2 Making Slides Hydrophilic Using Glow Discharge

1. Place plastic-coated slides in 25-slide plastic box and glow discharge for 10 min immediately before use. Detailed instructions for doing glow discharge are presented in (17).

2. Properly glow discharged slides will be hydrophilic and will remain usable for chromosome spreading for several hours. Be careful not to heat the slides, which can make it difficult to remove the plastic from the slide.

3.3 Preparing Protoplasts for Spreading SCs

1. Carefully pick 5–10 buds ~3 mm in length and place buds in a small beaker with a little distilled water.

2. Under a dissecting microscope, gently hold the apical end of the bud in place with the sharp dissecting needle, and use the chisel-shaped end of another dissecting needle to cut the bud open between two sepals to avoid damaging the anthers.

3. Gently pry open the bud, but do not touch the anthers.

4. Use the chisel-shaped dissecting needle to cut the filament at the base of one of the anthers, lift it out of the bud, and measure its entire length from base to tip using an ocular micrometer in a dissecting microscope. Typical anther lengths and stages are: leptotene—1.6 mm; zygotene—1.7 to 1.8 mm; pachytene—1.9 to 2.1 mm, but these can vary depending on the variety and health of the plant as well as the time of year. Keep the remaining anthers in the bud moist.

5. Determine the stage of meiosis by squashing the pollen mother cells (PMCs) from the anther in 2% aceto-orcein (17). Examine the slide at about 200–400× magnification, and if the primary microsporocytes are at the right stage, transfer the remaining four anthers into 0.2 ml of digestion medium in a depression slide.

6. Add 3 mg of desalted, lyophilized cytohelicase.

7. Using a small, sharp scalpel, slice the anthers transversely near the middle of their long axis.

8. Place the depression slide in a petri dish with water in the bottom, cover the petri dish, and wait 5 min for the cells to plasmolyze.

9. Under a dissecting microscope, use a chisel-shaped dissecting needle to press out the rods of cells from four half-anthers, and remove the anther walls.

10. Digest cells for 10–15 min at ~21 °C in the dark.

3.4 Spreading SCs

1. Draw (aspirate) two rods of cells in approximately 0.25 µl of the digestion medium into a siliconized micropipette. Keep the cell concentration high and use the lowest total volume of digestion medium possible (see Note 6). The bore of the micropipet tip should be just smaller than the diameter of a rod of cells, so that the rods break up but protoplasts are not damaged.

2. Gently expel the protoplast suspension into a 7 µl droplet of bursting medium suspended at the end of a 200 µl pipette tip.

3. Touch the droplet directly onto the surface of plastic-coated glow-discharged slide, and add another 7 µl of the bursting medium.

4. Take the slide immediately to a hood, and give it 30 sweeps of a fine mist of aqueous 4% formaldehyde using a nebulizer.

Blow the mist directly onto the slide about 1–2 cm from the mouth of the nebulizer. The frosted end of the slide should be visibly moistened during each pass, and the slide should not dry out during nebulization in spite of the air flow into the hood. The slide can be moved horizontally within the hood to make room for more slides, but do not pick the slide up until it is completely dry to avoid distorting the spreads.

5. Dry slides for at least an hour in the hood.

6. Use colored fingernail polish to paint a thin layer along the outer edges of each slide to prevent the plastic from separating prematurely from the glass. Allow fingernail polish to dry before proceeding.

7. Wash the slides once in aqueous 0.4% Photo-Flo 200 and twice in distilled water for 10–30 s in each wash. We use 3–150 ml beakers filled with 100 ml of solution and transfer the slides from one beaker to the next using forceps. Do not agitate the slides during washing. Wipe the back of the slides after the first wash to remove damaged plastic. If not, some of this plastic will lift off the back and may attach to the front of the slide and obscure SC spreads.

8. Air-dry the slides upright in a rack at room temperature. Stain the slides within a week. The slides can be left in the rack until staining, but protect them from dust.

3.5 DNase I Treatment, Fixation, and Post-staining for Electron Microscopy (See Note 7)

1. DNase I treatment (optional): dilute DNase I 1:1,000 in 1× DNase buffer, add 50–100 µl of solution to each slide, cover with a plastic coverslip (cut from an autoclave bag), and incubate at room temperature for 10 min. Add distilled water to edge of coverslip so that it floats above slide surface, and carefully remove coverslip with a pair of fine forceps. Gently wash the slide with a stream of water and air-dry. Reseal plastic edges with fingernail polish, if necessary. This optional DNase I treatment is done after the slides have been washed and air-dried (Subheading 3.4, steps 7 and 8 above) but before the fixation and staining steps described below.

2. Fix slides: in a hood, place about 200 µl of 2% FG on each slide, cover with a plastic coverslip, and incubate for 10 min at room temperature. Wash the plastic coverslip and fixative from the slide with distilled water, transfer the slide into a 150 ml beaker with 0.4% Photo-Flo 200 for about 30 s, then transfer the slide into a series of four 150 ml beakers containing distilled water for serial 30 s washes. After the first slide has been transferred to the second beaker, the same procedure of removing fixative and washing can be applied to the next slide. Do not agitate the slides in any wash steps. Place slides into a rack and air-dry overnight. Change all of the wash solutions after 10 slides have been washed.

3. PTA staining: place slides in a Coplin jar. Cover slides with freshly prepared alcoholic PTA stain and incubate 5–10 min. Remove one slide at a time from the stain with forceps and wash them with a gentle stream of 95% ethanol. Do not squirt the plastic directly or the plastic may lift up. Be sure to wash the back of the slide also to prevent precipitate from forming. Air-dry in a rack.

3.6 Scanning Slides for Good Spreads and Lifting Plastic from Slide

1. Scan each slide using phase-contrast microscopy at 200–400× to locate good spreads (Fig. 1a). The location of each spread on a slide should be recorded either using a vernier scale or a motor-driven stage. If there are many spreads, use a code to assist in later deciding which spreads to pick up onto grids (see Note 8).

2. Prepare tacky grids by dipping each grid into glue using a pair of EM forceps and then placing the grids onto waxed paper to dry. We always put the shiny side of each grid down onto plastic, so we put the shiny side of each dipped grid up to dry.

3. Select an SC spread by phase-contrast light microscopy, pick up a tacky grid, and gently place the grid onto the slide in the approximate location desired. Look through the oculars and see if the spread is visible within one of the grid openings. If not, move the grid around with the eyelash tool until it is visible in a grid opening. Continue with other grids. Often it is possible to arrange a grid so that several spreads are visible in different holes of the grid. Also, it may be necessary to move one grid slightly to allow another one to be placed close by.

4. Once all the grids have been positioned, carefully move the slide to a dissecting microscope. It is essential that the slide be held horizontal and not bumped to prevent the grids from moving.

5. Using a pointed steel probe, cut the plastic coating by carefully tracing completely around each grid leaving ~ a 1 mm margin. If necessary, trace around a group of closely spaced grids, leaving the same 1 mm margin on the edges. Trace around all grids and grid groups on a slide before going to the next step. Adjust the light on the dissecting microscope so that the edges of the newly cut plastic can be seen.

6. Place a small (~1–2 µl) drop of 1% HF near (but not on) each cut edge and draw the HF to the edge using a sharpened probe. The HF should immediately go under the plastic surface and begin to dissolve the glass so that the plastic begins to float above the glass. Once lifting has started, add 5% aqueous acetic acid to lift plastic and grids until they are floating completely free from the slide. We usually add HF to each plastic edge, and then quickly add the 5% acetic acid in the same order. Do

not leave the plastic and grids exposed to HF for more than a couple of min since a precipitate can form that interferes with EM visualization.

7. When all of the grids are freely floating above the slide, hold the slide level, and move it to an open glass bowl filled with distilled water. Gently push the slide into the water at a low angle. Once water starts coming over the slide surface, the plastic islands carrying grids will float free on the water surface, and then the slide can be carefully removed from the bowl (see Note 9).

8. Use a probe to "herd" the grids together into one part of the culture dish by gently moving the surrounding water and/or lightly touching the edges of the plastic.

9. Using a few centimeter wide piece of Parafilm, make contact with the floating grids, and pick them up with a smooth sweeping motion through the water. If we designate the "front" of the Parafilm as the side with plastic and grids and the back as the side without plastic or grids during the sweep, during the sweeping motion the water should continue to press on the "front" of the screen so that the plastic and grids stay in contact with the Parafilm. If the motion is not smooth, the grids and plastic may come loose on or in the water. If that happens, try to pick the grids up again with a fresh piece of Parafilm (see Note 10).

10. Once the grids are picked up onto the Parafilm, let the grids dry completely ($\geq$1 h).

11. Use EM forceps to remove each grid from the Parafilm. Place grids onto a clean microscope slide with plastic (shiny) sides facing up. Using phase-contrast microscopy, examine each grid and record the location of good spreads based on the center mark of the finder grid. Examine in the electron microscope (Figs. 1 and 2).

4 Notes

1. The plants must be in good health, and both buds and anthers must be treated gently (not bruised) to get good SC spreads.

2. All solutions should be made with high-quality, distilled water. Living cells are particularly sensitive to water quality, and we use only distilled water for those applications. Deionized water can be used for other, less sensitive applications such as slide washing.

3. For PTA staining, Falcon plastic works best. See ref. 17 for other plastic options and limitations.

4. Plastic surfaces are thin and fragile when wet or dry, so special care is required to avoid damage at all steps.

5. Because the atmosphere in the cylinder is saturated with dichloroethane, the plastic coating on the slide will continue to flow down the slide to make an ever thinner plastic coat. Adjust the time allowed for drawing the slide out of the cylinder to make minor changes in plastic thickness.

6. The best SC spreads are made using high cell concentration and little digestion medium. Before washing, the slides will have a thin, white coating of dried mannitol sugar. The presence of a large, dense, white mass of sugar on the slides shows that too much digestion medium was used to make the spreads.

7. All EM solutions need to be made using ultrapure distilled water.

8. Any spread that looks barely usable by light microscopy is unsuitable for EM.

9. We have not had any trouble with HF acid burns using these small volumes of dilute HF, but you may want to wear gloves during this operation since you may contact the HF containing water at some step. If you wear gloves, be sure that there is no powder residue that could contaminate the water surface and interfere with EM visualization.

10. While we have also used nylon to pick up grids, recently we found that grids are cleaner in the TEM when Parafilm is used.

Acknowledgments

This work was supported in part by National Science Foundation grants MCB-314644 and MCB-1019708- (L.K.A.) and DBI-0421634 (S.M.S.).

References

1. Fawcett DW (1956) The fine structure of chromosomes in the meiotic prophase of vertebrate spermatocytes. J Biophys Biochem Cytol 2: 403–406

2. Moses MJ (1956) Chromosomal structures in crayfish spermatocytes. J Biophys Biochem Cytol 2:215–218

3. Carpenter AT (1975) Electron microscopy of meiosis in Drosophila melanogaster females: II. The recombination nodule—a recombination-associated structure at pachytene? Proc Natl Acad Sci USA 72:3186–3189

4. Gillies CB (1973) Ultrastructural analysis of maize pachytene karyotypes by three dimensional reconstruction of the synaptonemal complexes. Chromosoma 43:145–176

5. Holm PB, Rasmussen SW (1980) Chromosome pairing, recombination nodules and chiasma formation in diploid *Bombyx* males. Carlsberg Res Comm 45:483–548

6. Counce SJ, Meyer GF (1973) Differentiation of the synaptonemal complex and the kinetochore in *Locusta* spermatocytes studied by whole mount electron microscopy. Chromosoma 44:231–253

7. Moses MJ, Slatton GH, Gambling TM, Starmer CF (1977) Synaptonemal complex karyotyping in spermatocytes of the Chinese hamster (*Cricetulus griseus*). III. Quantitative evaluation. Chromosoma 60:345–375

8. Gillies CB (1981) Electron microscopy of spread maize pachytene synaptonemal complexes. Chromosoma 83:575–591

9. Stack SM (1982) Two-dimensional spreads of synaptonemal complexes from solanaceous plants. I. The technique. Stain Technol 57:265–272

10. Albini SM, Jones GH (1984) Synaptonemal complex-associated centromeres and recombination nodules in plant meiocytes prepared by an improved surface-spreading technique. Exp Cell Res 155:588–592

11. Anderson LK, Stack SM (1988) Nodules associated with axial cores and synaptonemal complexes during zygotene in *Psilotum nudum*. Chromosoma 97:96–100

12. Anderson LK, Hooker KD, Stack SM (2001) The distribution of early recombination nodules on zygotene bivalents from plants. Genetics 159:1259–1269

13. Anderson LK, Doyle GG, Brigham B, Carter J, Hooker KD, Lai A et al (2003) High-resolution crossover maps for each bivalent of *Zea mays* using recombination nodules. Genetics 165:849–865

14. Stack SM, Royer SM, Shearer LA, Chang SB, Giovannoni JJ, Westfall DH et al (2009) Role of fluorescence in situ hybridization in sequencing the tomato genome. Cytogenet Genome Res 124:339–350

15. Lohmiller LD, De Muyt A, Howard B, Offenberg HH, Heyting C, Grelon M et al (2008) Cytological analysis of MRE11protein during early meiotic prophase I in Arabidopsis and tomato. Chromosoma 117:277–288

16. Qiao H, Lohmiller LD, Anderson LK (2011) Cohesin proteins load sequentially during prophase I in tomato primary microsporocytes. Chromosome Res 19:193–207

17. Stack SM, Anderson LK (2009) Electron microscopic immunogold localization of recombination-related proteins in spreads of synaptonemal complexes from tomato microsporocytes. Methods Mol Biol 558:147–169

Chapter 16

Analysis of the Synaptonemal Complex in *Brassica Using* TEM

Susan Armstrong

Abstract

Much of meiosis research is focussed on *Arabidopsis thaliana*, largely due to the significant advantages it brings, having a small sequenced genome with comparatively little repetitive DNA, the ease of forward and reverse genetics, and a short life cycle. On the other hand, due the small genome size using *Arabidopsis* may be problematic for generating sufficient meiotic material for other types of analysis e.g., proteomics using prophase meiocytes and cytological analysis of the synaptonemal complex at the subcellular level. One solution is to use closely related species with larger genomes, in this case the Brassicas. This chapter contains methods for spreading of *Brassica oleracea* meiocytes for the analysis of the synaptonemal complex by silver staining and immunolocalization with gold-coupled antibodies using transmission electron microscopy.

Keywords *Brassica oleracea*, Synaptonemal complex, Immunogold localization, Electron microscopy

1 Introduction

Arabidopsis thaliana and the crop genus *Brassica*, e.g., cabbages, oil seed rape, are members of the dicotyledonous family *Brassicaceae* (crucifers). Sequence comparisons suggest that *Arabidopsis* and *Brassica* diverged around 15–20 million years ago (1) after the last genome-wide duplication event in *Arabidopsis* (2). We have found that resources from *A. thaliana* can be exploited to investigate the chromosomal organization of its closest economically important relatives (3, 4). Plant meiosis research has focussed on *A. thaliana*, largely due to the significant genetic and resources available from this model dicotyledonous plant. On the other hand due the small genome size there are difficulties to generate sufficient meiotic material for other types of analysis, for example the organization of the prophase meiotic chromosomes at the ultrastructural level. This chapter contains methods for production of sections as well as spreading of *Brassica oleracea* meiocytes for the analysis of the

Wojciech P. Pawlowski et al. (eds.), *Plant Meiosis: Methods and Protocols*, Methods in Molecular Biology, vol. 990, DOI 10.1007/978-1-62703-333-6_16, © Springer Science+Business Media New York 2013

synaptonemal complex by silver staining and immunolocalization with gold-coupled antibodies using transmission electron microscopy.

---

2   Materials

2.1   Plants

We use plants of the *B. oleracea* var. *alboglabra* double haploid line A12DHd derived from *B. oleracea* var *alboglabra* by microspore culture (5). This line has the advantages of a short generation time (7–8 weeks), and is an erect robust plant unlike other rapidly cycling Brassicas. Sow *B. oleracea* seeds in 10 cm diameter pots in soil-based compost. Grow on the plants in dedicated chambers maintained at 18°C with supplementary lighting when necessary, in a 16-h day length.

2.2   Sectioning Meiocytes

1. Fixative: 2% glutaraldehyde, 2% paraformaldehyde both EM grade and 0.1 M sodium cacodylate pH7.0 in sterile deionized water. Make freshly before use.

2. Alcohol series: 70, 85, 90% of absolute ethanol in sterile deionized water.

3. 50% LR White: mix 1 volume LR White (Agar Scientific, Stansted, UK) with 1 volume of 90% ethanol.

4. 100% LR White.

5. Gelatine capsules (Agar Scientific, Stansted, UK).

6. Formvar coated grids: for standard EM copper grids may be used. For immunolocalization we use nickel grids. Make formvar solution by dissolving 0.75 g of formvar in 100 ml chloroform. Glass slides are dipped in the formvar solution, air-dried and the plastic is floated on a large volume of water. Using a pair of fine forceps the grids are placed on the plastic, and this is removed from the water using paper, the grids are air dried and stored in a petri dish.

7. Staining solution for sections: dissolve either 30% uranyl acetate in methanol or 2% uranyl acetate in water.

2.3   Spreading Brassica SCs

1. Enzyme digestion medium: dissolve 0.1 g (for final concentration of 0.4%) cytohelicase (Sigma Aldrich, St. Louis, MO, USA) with 0.375 g (for final concentration of 1.5%) sucrose and 0.25 g (for final concentration of 1%) polylvinylpyrrolidone (MW 40,000; Sigma Aldrich, St. Louis, MO, USA) in 25 ml sterile deionized water. Dispense 1 ml aliquots and store at −20°C.

2. Spreading medium: 0.05% Triton X-100 in freshly distilled water.

3. Paraformaldehyde fixative: weigh out 4 g of paraformaldehyde (electron microscopy grade) in the fume hood. Dissolve in

100 ml of sterile deionized water and 4 drops of 1 M NaOH, pre-warmed to 60°C in the microwave. Stir the mixture on a magnetic stirrer for 1 h, or until it dissolves. Filter through Whatman paper. Adjust the pH to 8.0. The fixative can be stored for up to 1 week at 4°C.

4. Plastic coated slides: dip the slides in a solution comprising 0.75 g petri dishes (Sterilin, Stafford, UK) cut up and dissolved in 100 ml chloroform. The solution is satisfactory for 1 month, after which it needs to be replaced. The slides are glow discharged in order to remove static charge.

5. Make up a 50% silver nitrate solution for staining of axial/lateral elements immediately before using.

6. Cut-up rectangles of nylon gauze to cover the slides.

7. Finder grids (Agar Scientific, Stansted, UK). For analysis of spreads we use copper finder grids to facilitate their location with EM.

2.4 EM Immunocytology

1. Blocking buffer: dissolve 1% Bovine Serum Albumin (BSA) in Phosphate buffer saline, PBS, (we tend to use prepared tablets, but PBS can also be made up from a 10× stock containing 1.37 M NaCl, 27 mM KCl, 100 m Na_2HPO_4, 18 mM KH_2PO_4, adjusted to pH 7.4 if necessary). Autoclave and store at room temperature. The working solution of the buffer is 1: 10 with sterile deionized water.

2. Primary antibodies: make up to the relevant dilution, e.g., 1:50, 1:100, 1:500, 1:1,000 in 1% BSA and PBS with 0.1% Triton X.

3. Washing solution: PBS + 0.1% Triton X-100.

4. Secondary antibodies: make up at 1:50 using conjugates of 5 nm gold or 10 nm gold, (BB International, Cardiff, UK).

5. Staining solution: Dissolve either 30% uranyl acetate in methanol or 2% uranyl acetate in water.

3 Methods

Electron microscopy studies of meiosis in Brassicas have allowed us to analyze the pairing and synapsis of homologous chromosomes at an ultrastructural level (Figs. 1 and 2). We also have developed techniques for immunolocalization (3) (Fig. 3).

3.1 Sectioning Brassica Meiocytes for EM

1. Take young terminal inflorescences from the plants and place them on damp filter paper in a petri dish. Dissect out individual buds, size range 0.5–2 mm, using a stereomicroscope and order them by size. Remove single anthers from each bud,

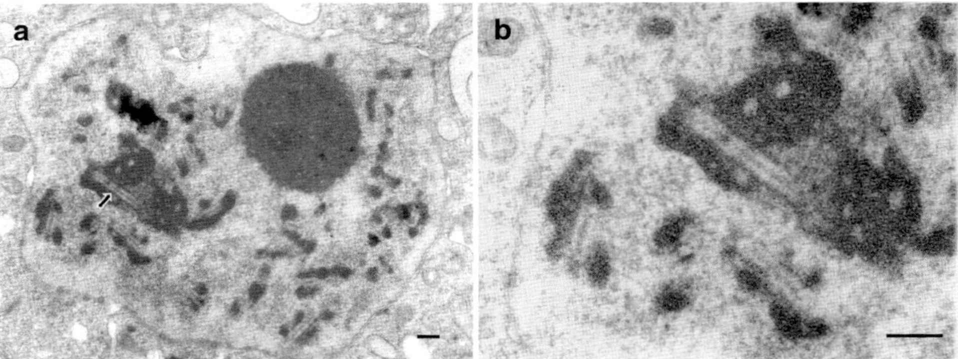

Fig. 1 Ultrathin section of *Brassica* anther at pachytene. Brassica anthers were embedded in LR White and stained with uranyl acetate. (**a**) A whole nucleus. *Arrow* indicates a fragment of the SC. (**b**) Enlargement of the area containing the SC fragment. Bar = 500 nm

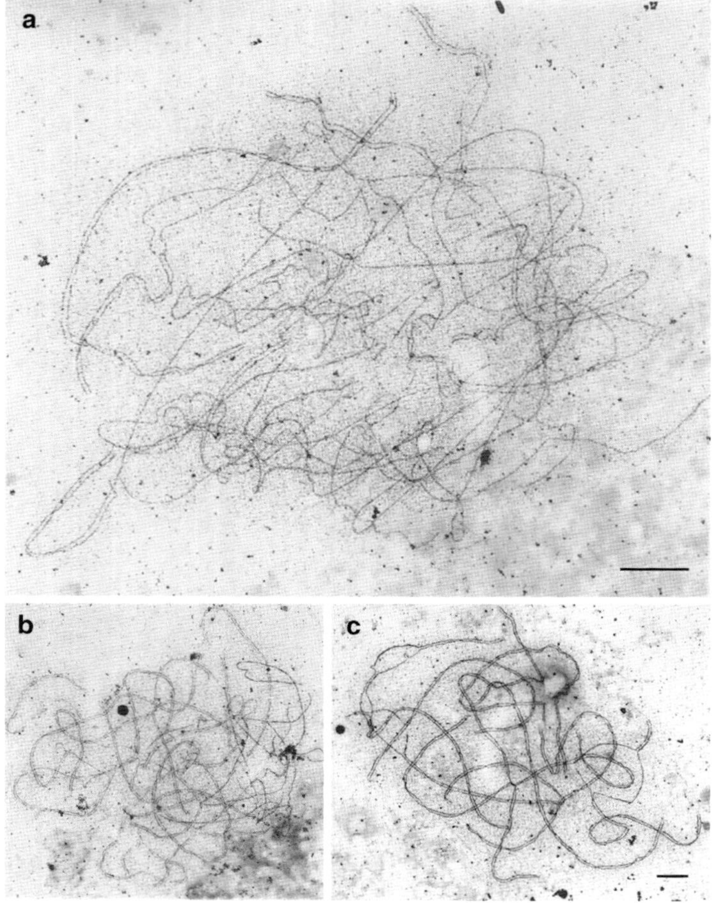

Fig. 2 Spread *Brassica* SC stained with silver nitrate. (**a**) Early zygotene. (**b**) Late zygotene. (**c**) A cell with nearly completed synapsis. Bar = 2 μm

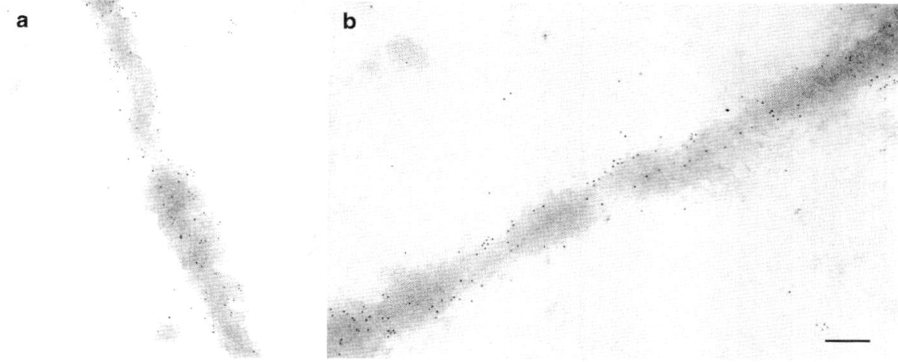

Fig. 3 Immunolocalization of ZYP1 on *Brassica* SC. Two regions of SC in a pachytene meiocyte of *Brassica* immunolabeled with an antibody against the C terminus portion of ZYP1 produced in rat (6) and detected with a secondary antibody coupled to 10 nm gold particles. Bar = 200 nm

place them on microscope slides and squash in lacto-propionic stain to determine their meiotic stage (see Note 1). Collect the remaining anthers (up to 5) in the bud, which will be at a similar stage (e.g., pachytene for analysis of the mature synaptonemal complex). Fix the anthers individually overnight.

2. Dehydrate through alcohol series to 90% ethanol.

3. Place in 1:1 ratio of 90% ethanol:LR White for 2 h.

4. Replace with the same solution for a further 2 h.

5. Incubate 100% LR White overnight, up to 48 h.

6. Place specimen in gelatine capsules and out-gas.

7. Polymerize under UV light for 2–3 days.

8. Cut semi-thin sections onto a clean water surface and pick them up onto formvar-coated copper grids that are, in turn, picked onto clean paper and air-dried.

9. Stain with 30% uranyl acetate in methanol for 7 min or 2% uranyl acetate in water for 1 h at 40°C.

10. Examine grids in a JEOL 1200 EX electron microscope (Fig. 1).

3.2 Spreading Brassica SC for E.M.

1. Take young terminal inflorescences from the plants and place them on damp filter paper in a petri dish. Dissect out individual buds, size range 0.5–2 mm, using a stereomicroscope and order them by size. Remove single anthers from each bud, place them on microscope slides and squash in lacto-propionic stain to determine their meiotic stage (see Note 1). The remaining anthers (up to 5) in the bud will be at a similar stage. Collect a pool of around 10 anthers at a similar stage (e.g., pachytene for analysis of the mature synaptonemal complex).

2. Place around 10 anthers together in a cavity slide containing 15 µl digestion medium and leave them for 4 min in a moisture box at 33°C.

3. Tap the anthers gently with a brass rod to release meiocytes. It may be necessary to add more digestion medium in order to avoid drying out of the suspension. Place in a moist box for a further 4 min including tapping out.

4. Place 2 µl of this solution onto 10 µl drops of spreading medium on glow discharged plastic coated slides.

5. Leave the suspension for a further 2 min, and monitor with a phase contrast microscope until the meiocytes can be seen to be bursting.

6. Spread the suspension over the slide using a spreader made from a glass pipette and fix the cells by adding 20 µl of paraformaldehyde fixative in the fume hood. Allow to dry.

7. Wash the dry slides in sterile deionized water and stain using silver nitrate. Place around 200 µl of silver nitrate solution on the slides and cover them with nylon mesh. Incubate in a moist box for 20–30 min at 60°C until the mesh appears to be golden brown. Wash the slides and nylon mesh in a large volume of water, the nylon mesh will float away. Rinse the slides in sterile deionized water and dry.

8. Examine the stained slides with a light microscope and mark well spread cells with black ink before floating the plastic supporting film onto a clean water surface. To do this, score the area of interest is with a razor blade and gently float it off onto the water.

9. Place copper finder grids onto the marked area and pick up the plastic film on clean paper. Air-dry. Score the plastic area around the grid and store individual grids.

10. Examine grids in a JEOL 1200 EX electron microscope (Fig. 2).

3.3 Immunolocalization on Sections

1. Prepare the sections as already described to the stage before staining (see Subheading 3.1, steps 1–7).

2. Cut semi-thin sections and float the supporting film onto clean water surface (see Subheading 3.1, step 8). For immunolocalization, use formvar-coated nickel grids. Note that spreads will be on dull sides.

3. Block nonspecific binding by incubation with a blocking buffer. Place grid, dull side down, into a drop of about 200 µl of block in petri dish or placed on dental wax.

4. Transfer grids to primary antibody and incubate overnight at 4°C in a petri dish.

5. Wash three times for 5 min in washing buffer by placing the grids sequentially into 3 drops of about 200 µl washing buffer.

6. Place the grid dull side down into a drop of secondary antibody for 90 min at room temperature.

7. Wash three times for 5 min in washing buffer.

8. Wash three times for 1 min in sterile deionized water.

9. Stain with 30% uranyl acetate in methanol for 7 min or 2% uranyl acetate in water for 1 h at 40°C.

10. Examine grids in a JEOL 1200 EX electron microscope.

3.4 Immunolocalization for Spreads

1. Prepare the spreads as already described, to the stage before silver staining (see Subheading 3.2, steps 1–7).

2. Examine the slides with a phase-contrast microscope, mark well spread cells with black ink before floating the plastic supporting film onto a clean water surface

3. Place formvar-coated nickel finder grids onto the marked area and pick up the plastic film on clean paper. Air-dry. Note that spreads will be on dull sides.

4. Block nonspecific binding by incubation with a blocking buffer. Place grid, dull side down, into a drop of about 200 µl of block in petri dish or on dental wax.

5. Transfer grids to primary antibody and incubate overnight at 4°C.

6. Wash three times for 5 min in washing buffer.

7. Apply secondary antibody for 90 min at room temperature.

8. Wash three times for 5 min in washing buffer.

9. Wash three times for 1 min in sterile deionized water.

10. Stain with 30% uranyl acetate in methanol for 7 min or 2% uranyl acetate in water for 1 h at 40°C.

11. Examine grids in a JEOL 1200 EX electron microscope.

4 Note

1. We use the stain lactopropionic orcein to check the meiotic stage of meiocytes in *Brassica*. To make, dissolve 5 g orcein in 100 ml of 1:1 mix of lactic acid and propionic acid. Leave stirring overnight in fume hood at room temperature to dissolve completely. The next day, filter through Whatman No 1 filter paper. This is the stock. The working solution is a 1:1 dilution with distilled water.

Acknowledgments

The *Brassica* work in our laboratory is supported by the Biotechnology and Biological Sciences Research Council (BBSRC). Technical support has been provided by Steve Price.

References

1. Yang Y-W, Lai K-N, Tai P-Y, Li W-H (1999) Rates of nucleotide substitution in angiosperm mitochondrial DNA sequences and dates of divergence between *Brassica* and other angiosperm lineages. J Mol Evol 48:597–604

2. Bowers JE, Chapman BA, Rong J, Paterson AH (2003) Unravelling angiosperm genome evolution by phylogenetic analysis of chromosomal duplication events. Nature 422:433–438

3. Armstrong SJ, Caryl AP, Jones GH, Franklin FCH (2002) Asy1, a protein required for meiotic chromosome synapsis, localizes to axis-associated chromatin in Arabidopsis and Brassica. J Cell Sci 115:3645–3655

4. Sanchez-Moran E, Mercier R, Higgins JD, Armstrong SJ, Jones GH, Franklin FCH (2005) A strategy to investigate the plant meiotic proteome. Cytogenet Genome Res 109: 181–189

5. Bohuon EJR, Keith DJ, Parkin IAP, Sharpe AG, Lydiate DJ (1996) Alignment of the conserved C genomes of *Brassica oleracea* and *Brassica napus*. Theor Appl Genet 93:833–839

6. Higgins JD, Sanchez-Moran E, Armstrong SJ, Jones GH, Franklin FCH (2005) The *Arabidopsis* synaptonemal complex protein ZYP1 is required for chromosome synapsis and normal fidelity of crossing over. Genes Dev 19:2488–2500

Chapter 17

Preparing Thin Sections of Meiotic Nuclei for Transmission Electron Microscopy

Ljudmilla Timofejeva

Abstract

Transmission electron microscopy (TEM) of thin sections of meiocytes embedded in plastic resin allows analysis of structures in meiotic nuclei, such as the synaptonemal complex, at very high resolution while preserving the spatial organization of chromosomes.

Keywords Transmission electron microscopy, Thin sections, Meiosis, Synaptonemal complex (SC), Synapsis, Anther, *Arabidopsis thaliana*

1 Introduction

The reductional division of meiosis includes homologous chromosome pairing, synapsis, and recombination. Synapsis results in the formation of the synaptonemal complex (SC), a proteinaceous structure that connects homologous chromosomes. At leptotene, sister chromatids organize along structures called axial elements (AEs). During zygotene, transverse filaments connect the AEs to form the tripartite SC (1). The SC structure can be observed using transmission electron microscopy (TEM). In addition, electron-dense bodies called recombination nodules, which are associated with the SC, can be detected using TEM (2) (Fig. 1).

In flowering plants, stamens and carpels are the male and female reproductive organs, which produce meiotic cells. A stamen consists of a four-lobed anther supported by a filament. During anther development, four layers of somatic cells differentiate inside each lobe around germinal cells competent for meiosis. Development of these distinctive cell types is highly coordinated and meiotic stages can be precisely assigned, even in meiotic mutants with disrupted chromosome pairing and/or synapsis, based on the developmental

Wojciech P. Pawlowski et al. (eds.), *Plant Meiosis: Methods and Protocols*, Methods in Molecular Biology, vol. 990, DOI 10.1007/978-1-62703-333-6_17, © Springer Science+Business Media New York 2013

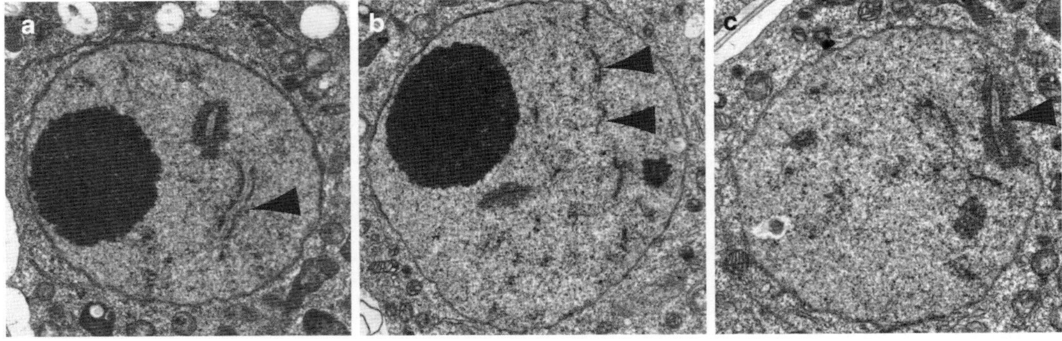

Fig. 1 Thin sections of *Arabidopsis* meiocytes visualized using transmission electron microscopy. (**a**) Synaptonemal complex (SC) fragments (one marked by *arrowhead*) in a wild-type pachytene nucleus. (**b**) Unsynapsed fragments of axial elements (two examples marked by *arrowheads*) in a pachytene nucleus of the *atxrcc3* mutant (3). (**c**) A rare SC fragment (marked by *arrowhead*) in a pachytene nucleus of the *atxrcc3* mutant (3). Bar = 2 μm

stage of the anther wall cells surrounding the meiocytes. Thus, TEM analysis of anther cross sections allows distinguishing meiotic stages even in the absence of specific features of these stages (presence of AEs at leptotene, progression of synapsis at zygotene, and complete SC formation at pachytene).

There are important differences between animal and plant cells with respect to tissues preparation for electron microscopy. These differences include the presence of the cell wall and vacuoles, as well as low concentration of proteins in plant cells. The thick callose wall surrounding the meiotic cells tends to resist fixative penetration. Therefore, for satisfactory fixation of plant meiocytes, special treatments are required (4). These special treatments include adding Triton X-100 to the fixative to permeabilize cell membranes, carrying out fixation with glutaraldehyde in N-2-hydroxyethylpiperazine-N'-2-ethane sulfonic acid (HEPES) buffer, and performing prolonged tissue infiltration with Spurr's low viscosity embedding mixture consisting of a cycloaliphatic epoxide resin, diglycidyl ether of polypropylene glycol as a flexibilizer to control the hardness of the polymerized block, nonenyl succinic anhydride as a hardener, and dimethylaminoethanol as an accelerator (5). This resin provides excellent penetration and rapid infiltration of plants tissues. The blocks have good trimming and sectioning qualities and the sections are tough under the electron beam.

2   Materials

2.1   Plants

Grow *Arabidopsis thaliana* (see Note 1) wild-type and mutant plants in Metro-Mix 200 soil (Scotts-Sierra Horticultural Products Co.,

Maryville, OH, USA) or similar in a growth room or greenhouse at the temperature of 18–22 °C under long-day conditions (16 h day and 8 h night).

2.2 Reagents

Use ultrapure water for preparing all solutions.

1. 2× HEPES buffer: 0.2 M HEPES, 0.04% (v/v) Triton X-100, pH 7.2 (see Note 2). Weigh 5.2 g of HEPES sodium salt and transfer to a glass beaker. Alternatively, 4.77 g of HEPES free acid can be used, although the sodium salt of HEPES is more soluble than its free acid counterpart. Add water to a volume of 80 ml, dissolve HEPES, and add 200 μl of 20% Triton-X 100. Mix well and adjust pH with NaOH to 7.2 (see Note 3). Make up to 100 ml with water. Concentrated HEPES buffer can be prepared in advance and stored at 4 °C for up to a month. It should be mixed with the fixative just before use.

2. Glutaraldehyde fixative: 2.8% glutaraldehyde in 1× HEPES buffer. Prepare by mixing 2.8 ml of 25% glutaraldehyde with 12.5 ml of 2× HEPES buffer containing 0.04% Triton X-100, make up to 25 ml with water (see Note 4). Prepare only the needed amount prior to use and do not store.

3. 1% osmium tetroxide (OsO_4) (see Note 5): dissolve OsO_4 crystals in water in a dark glass bottle (see Note 6) to make a 4% aqueous solution. Store tightly closed in a cold place. Several hours are required to completely dissolve OsO_4 crystals. Dilute 4% OsO_4 to make a 2% solution by mixing 1 volume of OsO_4 stock solution with an equal volume of water, and then add two volumes of the 2× HEPES buffer. Osmium tetroxide is extremely toxic even at low concentrations and low exposure levels and must be handled with appropriate precautions. Wear gloves and perform all work in a fume hood. Neutralize used fixative with corn oil. Osmium tetroxide waste should be collected in a bottle for safe disposal.

4. A series of ethanol solutions for tissue dehydration starting with 30% ethanol to 100% ethanol in 10% increment (see Note 7).

5. 1:1 Ethanol:acetone (v/v). Use 100% ethanol to make the solution.

6. Spurr's embedding kit (Electron Microscopy Sciences, Hatfield, PA, USA). To prepare the resin, follow manufacturer's recommendations. Add components one by one while mixing the solution using a magnetic stirrer. The accelerator should be added last and well mixed. Unused resin mix can be stored for a couple of days in plastic syringes at 4 °C.

7. A series of Spurr's resin solutions in acetone: 1:3, 1:2, 1:1, 2:1, and 3:1 (v/v).

8. 0.1% Toluidine blue (w/v) (Electron Microscopy Sciences, Hatfield, PA, USA) in 1% sodium tertraborate (Electron Microscopy Sciences, Hatfield, PA, USA).

9. Saturated uranyl acetate ($UO_2(CH_3COO)_2 \cdot 2H_2O$) in ethanol (see Note 8). Uranyl acetate is photo-unstable. To prevent precipitation, wrap the tube in foil.

10. CO_2-free water: bring double distilled water to boiling and then shake it vigorously for at least 1 min.

11. Lead citrate solution ($Pb(C_6H_5O_7)_2 \cdot 3H_2O$). Mix 1.33 g of lead nitrate ($Pb(NO_3)_2$), 1.76 g of sodium citrate ($Na_3(C_6H_5O_7) \cdot 2H_2O$), and 30 ml of CO_2-free water. Add 8 ml of 1 N NaOH and mix until the solution becomes clear. The pH should be 12.0 ± 0.1. If it is lower, add more NaOH. If it is higher, start over, this time adding a smaller amount of NaOH. Bring the solution to a final volume of 50 ml by adding CO_2-free water. Let stand for several hours before use. Store in a tightly closed bottle at 4°C.

2.3 Equipment and Supplies

Specialized items can be purchased from electron microscopy suppliers.

1. Small glass vials with plastic caps.

2. Vial rack.

3. Plastic transfer pipette.

4. Flat bullet-shaped indexed embedding molds.

5. Disposable plastic beakers.

6. Controlled-temperature oven for resin polymerization.

7. Razor blades for resin block trimming.

8. Ultramicrotome with a diamond knife for ultrathin sectioning.

9. Glass knife boats.

10. Glass strips for making glass knives.

11. Sheets of dental wax.

12. Nonmagnetic stainless tweezers with very fine tips.

13. Copper grids with standard square or hexagonal mesh.

14. A grid storage box.

15. A perfect loop to place ultrathin sections on grid mesh (optional).

16. Glass microscope slides.

17. A silicone staining pad or the EMS staining plate (Electron Microscopy Sciences, Hatfield, PA, USA) with wells for staining sections on grids.

3   Methods

3.1 Fixation of Arabidopsis Inflorescences in Glutaraldehyde Fixative

1. Prepare numbered Eppendorf tubes with 500–700 µl of fixative solution.

2. Select an inflorescence with one or two open flowers, dissect a small part of the inflorescence using tweezers, and place it in the fixative (see Note 9). Submerge the inflorescence immediately and push the sample to the bottom of the tube with a cut pipette tip to remove air bubbles. Air bubbles inside flower buds will prevent tissue penetration by the fixative and will cause the tissue to float to the surface of the fixative.

3. Let the fixative penetrate the tissue for 2–4 h at room temperature, then exchange the fixative and continue fixation at 4°C overnight.

4. Remove the fixative with a plastic transfer pipette and wash the samples in 1× HEPES buffer for 20 min.

5. Remove the washing buffer and transfer samples to glass vials filled with 1% OsO_4 in 1× HEPES buffer. Place the vials in a dark container and leave for 2–4 h at 4°C.

6. Rinse samples several times, 20 min. each time, in 1× HEPES buffer.

7. Dehydrate the samples in the series of ethanol solutions starting from 30% ethanol. Keep samples for 20 min in each solution on a slow shaker. Samples can be left in 70% ethanol overnight at 4°C if necessary. After reaching 100% ethanol, transfer samples to the ethanol:acetone (1:1) mix and complete the dehydration process with two washes in pure acetone.

3.2 Embedding

1. Infiltrate samples with Spurr's resin (see Note 10) starting with incubation in the 1:3 mix of Spurr's resin and acetone for 1–2 h. Continue infiltration by increasing the concentration of Spurr's resin in the mix: move the sample for 2–4 h to the 1:2 Spurr's resin: acetone mix and then for 4–8 h to the 1:1 mix (see Note 11). Leave the samples overnight in the 2:1 mix with shaking. The following day, incubate samples in the 3:1 mix for several hours and transfer them to pure resin for overnight incubation. Then, exchange the resin for a fresh one and incubate for the whole day.

2. Carefully dissect anthers out of the buds in a small petri dish using toothpicks. Anthers should remain immerged in Spurr's resin and should not be allowed to dry. Do not use sharp needles as they can damage the fragile tissue.

3. Place a paper label at the bottom of the mold and fill the mold halfway up with fresh Spurr's resin. Transfer dissected anthers or buds from the petri dish to the mold and orient them for

cross (transverse) or longitudinal sectioning. Using a plastic transfer pipette, fill up the mold with resin. Add resin all the way to the top of the mold as the volume of the resin will decrease after polymerization. Place molds with samples in an oven set at 60°C and allow the resin to polymerize for 2 days. After polymerization, blocks with embedded samples can be used immediately or can be stored for a long time without any loss of quality.

3.3 Sectioning

1. Put the resin block in a block holder and trim it with a razor blade to obtain a shape of a trapezoid on the front surface of the block.

2. Perform thin sectioning using an ultramicrotome. Diamond knives are the best choice for making thin sections. Alternatively, glass knives made out of glass strips can be used. Collect the sections on water. If sections are wrinkled, hold a wooden rod wetted with chloroform above the section (but do not touch the water surface) to smooth it.

3. Pick sections using the perfect loop and place them onto section grids. Alternatively collect sections directly onto a grid by touching the sections floating on the surface of the water with the grid.

4. Before staining, the grids with sections should be dried up for at least 2 h and ideally overnight.

3.4 Staining

Double staining with uranyl acetate and lead citrate is a common method for staining EM specimen.

1. Before starting the staining procedure, filter both uranyl acetate and lead citrate through 1 ml syringe filters to remove any precipitates.

2. Set up the staining apparatus: place a sheet of dental wax in a petri dish that will serve as the staining chamber and put several pellets of NaOH around it. NaOH will quickly absorb CO_2, minimizing lead citrate precipitation (see Note 12). These preparations should be made before the grids are stained with uranyl acetate so that the atmosphere in the chamber is free of CO_2 by the time of lead citrate staining.

3. Place several drops of a saturated solution of uranyl acetate on a clean sheet of dental wax in the petri dish.

4. Place the grid with the sections facing down on the uranyl acetate drops and allow the grid to float on the surface of the drops for 5–15 min. Rinse the grid twice with CO_2-free water. To wash the grid, hold it by its edge with forceps and dip rapidly in water in a small beaker. Use forceps with thin and fine tips to handle grids. Negative action tweezers are best for this

purpose as they allow holding the grid without applying finger pressure. Do not let the grids dry before continuing to lead citrate staining (6).

5. Place the grid with the sections facing down on a drop of lead citrate and stain for 3–15 min. Cover the dish immediately after placing the grid.

6. Wash the grid quickly in 0.02 N NaOH solution and then in distilled water. Make sure that all stain is washed off from the grid and the tips of the forceps. Any leftover stain between the tips of the forceps is a potential source of contamination. Such contamination can be avoided by placing a piece of filter paper between the tips of the forceps.

7. Allow the washed grid to dry.

3.5 Microscopy

Analyze grids with the Transmission Electron Microscope using an accelerating voltage of 80 kV. If contrast in your sections is very low or the sections are very thin, use the smallest objective aperture and an accelerating voltage of 60 kV instead of 80 kV to enhance the contrast.

4 Notes

1. I have used this protocol with good results with *Arabidopsis* ecotypes Landsberg *erecta* (L*er*), Columbia, and Wassilewskija.

2. The HEPES buffer is used to maintain the pH of the fixative. Alternatively, 0.1 M phosphate buffer or sodium cacodylate buffer can be used. However, HEPES is a zwitterionic buffer ideally suited for plant cells. It is not as toxic as the cacodylate buffer and works well in electron microscopy procedures.

3. The range of pH in the fixative is critical for chromatin structure preservation and should be between 7.2 and 7.4.

4. Use EM grade glutaraldehyde. Contamination with polymerized glutaraldehyde greatly reduces glutaraldehyde efficiency as a cross-linking agent and prevents proper fixation. For this reason, it is better to purchase single-dose ampoules sealed under dry nitrogen and keep them refrigerated.

5. Osmium tetroxide is used in TEM to provide contrast in fixed biological samples. Because OsO_4 molecules are nonpolar, they easily penetrate charged cell membranes and stabilize many proteins without destroying their structural features. Proteins stabilized by OsO_4 do not become coagulated by alcohols during the subsequent dehydration step (7). The osmium treatment embeds the heavy metal into cellular structures,

particularly cell membranes, blocking the electron beam and creating contrast with the neighboring cytoplasm and karyoplasm. Another heavy metal, i.e., uranium, can also be used as a contrast agent. Uranyl acetate binds to nucleic acids and proteins (8).

6. Osmium tetroxide can penetrate plastics and therefore should be stored in a glass vial.

7. Using lower increments (such as 10%) in the ethanol series, is better as it reduces tissue shrinkage.

8. Uranyl acetate is toxic and slightly radioactive. The amount of emitted radioactivity depends on which specific uranium isotopes are present in the reagent stock. Commercial stocks prepared from depleted uranium typically exhibit radioactivity in the range of 0.37–0.51 mCi/g.

9. Tissue pieces should not be larger than 1–1.5 mm3.

10. Many of the resin components are known to be carcinogenic.

11. Keep the samples in the 1:1 mixture of Spurr's resin and acetone for at least 4 h. Failure to increase the infiltration times when moving samples to the next mixture may result in poor infiltration. Poorly infiltrated sections will fail under the electron beam.

12. The lead citrate stain precipitates quickly upon contact with CO_2 (6).

References

1. Ma H (2006) A molecular portrait of Arabidopsis meiosis. The Arabidopsis book 4:e0095.

2. Anderson LK, Hooker KD, Stack SM (2001) The distribution of early recombination nodules on zygotene bivalents from plants. Genetics 159:1259–1269

3. Bleuyard JY, White CI (2004) The *Arabidopsis* homologue of Xrcc3 plays an essential role in meiosis. EMBO J 23:439–449

4. Owen HA, Makaroff CA (1995) Ultrastructure of microsporogenesis and microgametogenesis in *Arabidopsis thaliana* (L.) *heynh* ecotype Wassilevskija (*Brassicaceae*). Protoplasma 185:7–21

5. Spurr AR (1969) A low-viscosity epoxy resin embedding medium for electron microscopy. J Ultrastruct Res 26:31–42

6. Hayat MA (2000) Principles and techniques of electron microscopy: biological applications. Cambridge University Press, Cambridge, UK

7. Cooper K (1988) Neutralization of osmium tetroxide in case of accidental spillage and for disposal. Bull Microscopical Soc Canada 8:24–28

8. Bozzola JJ, Russell LD (1992) Electron microscopy principles and techniques for biologists. John and Bartlett Publishers, Boston, USA

Part III

Genetics and Molecular Biology Techniques

Characterization of Meiotic Non-crossover Molecules from *Arabidopsis thaliana* Pollen

Hossein Khademian, Laurène Giraut, Jan Drouaud, and Christine Mézard

Abstract

Meiotic recombination is essential for proper segregation of homologous chromosomes and thus for formation of viable gametes. Recombination generates either crossovers (COs), which are reciprocal exchanges between chromosome segments, or gene conversion not associated with crossovers (NCOs). Both kinds of events occur in narrow regions (less than 10 kb) called hotspots, which are distributed along chromosomes. While NCOs may represent a large fraction of meiotic recombination events in plants, as in many other higher eukaryotes, they have been poorly characterized due to the technical difficulty of detecting them. Here, we present a powerful approach, based on allele-specific PCR amplification of single molecules from pollen genomic DNA, allowing detection, quantification and characterization of NCO events arising at low frequencies at recombination hotspots.

Keywords Meiosis, Recombination, Hotspot, Non-crossover, Pollen, Allele-specific PCR

1 Introduction

In most eukaryotes, recombination between homologous chromosomes is essential for proper chromosome segregation at the first meiotic division. Meiotic recombination produces either reciprocal exchanges of large fragments (crossovers, COs) or non-reciprocal exchanges of small patches (non-crossovers, NCOs) of genetic material (Fig. 1). Recombination events are not distributed homogeneously along chromosomes but, instead, cluster at "hotspots," small regions a few kilobase-long with a higher CO and NCO rates compared to the surrounding DNA. Studies in yeasts have shown that both COs and NCOs derive from repair of DNA double strand breaks induced by the Spo11 protein early in meiosis (reviewed in ref. 1).

Wojciech P. Pawlowski et al. (eds.), *Plant Meiosis: Methods and Protocols*, Methods in Molecular Biology, vol. 990, DOI 10.1007/978-1-62703-333-6_18, © Springer Science+Business Media New York 2013

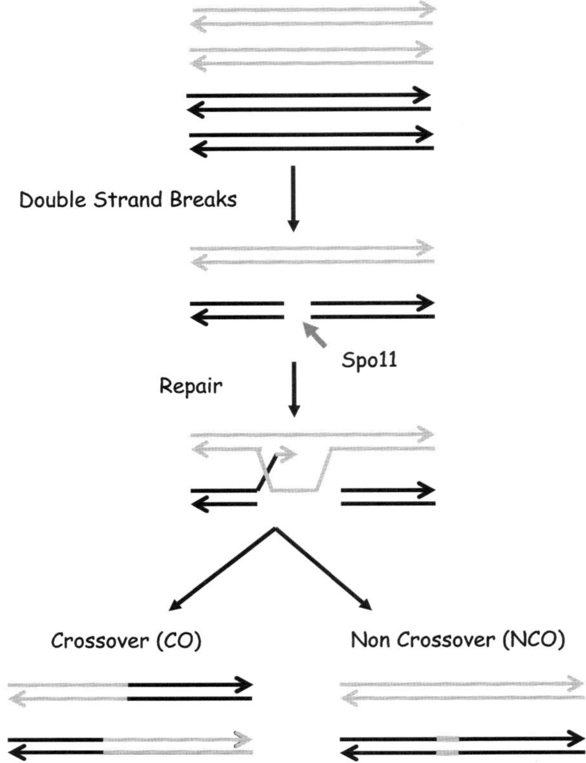

Fig. 1 Schematic cartoon of CO and NCO formation during meiosis. In early prophase I of meiosis, recombination is initiated by programmed DNA double strand break (DSB) formation by the Spo11 protein. Repair of these DSBs using the homologous chromosome as a template leads to the formation of crossover and non-crossover

In higher eukaryotes, the sperm typing technique set up first by Arnheim and Jeffreys enabled characterization of tens of hotspots in human and mice (2, 3) (Fig. 2). Mammalian hotspots exhibit similar characteristics to the ones described in yeasts: a 1–2 kb width, a quasi normal distribution of CO exchange points, and a gradient of the NCO rate peaking at the center of the hotspot.

In all studies conducted in higher eukaryotes, the number of NCOs is probably underestimated because of the technical difficulty of their detection and the limited number of available polymorphisms that enable genetic detection of recombination events. However, the numbers of DSB repair sites estimated by cytological methods (immunocytology using antibodies against the DSB repair protein RAD51 and/or DMC1, or quantification of early recombination nodules visualized by electron microscopy) are 10–40 times greater than CO numbers (estimated by chiasma count, quantification of late recombination nodules visualized by electron microscopy, or by genetic mapping) (1, 4). It is not known yet if

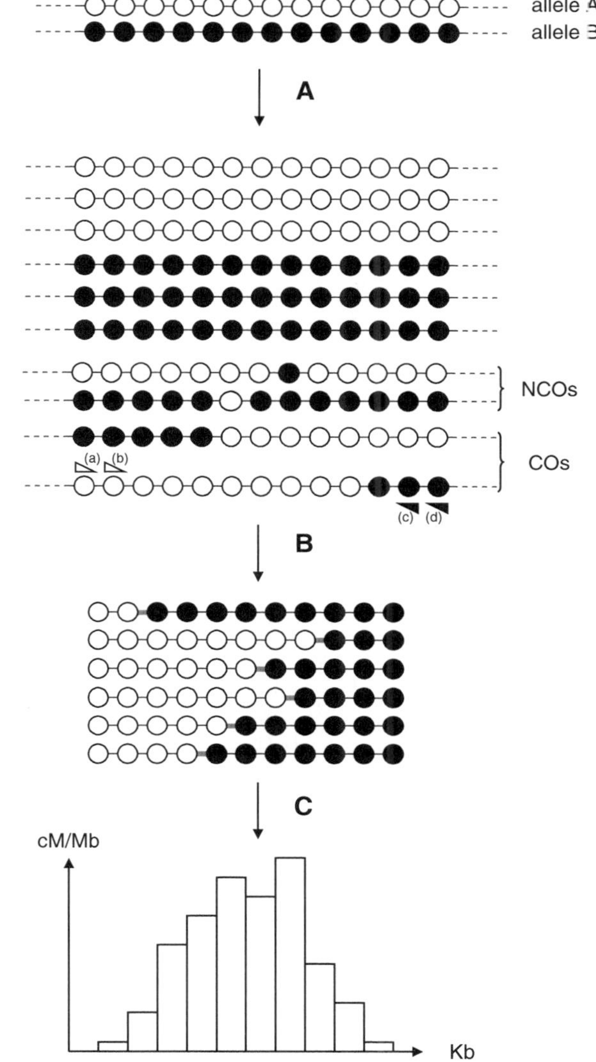

Fig. 2 Outline of the sperm typing and pollen typing assays. (**A**) Genomic DNA is extracted from sperm or pollen of an F1 (A × B) hybrid. At a given hotspot locus, the majority of the molecules contains either the A or the B allele. Only a few molecules contain crossover or non-crossover recombination products. (**B**) Crossover molecules are specifically amplified using two rounds of PCR with Allele Specific Oligonucleotides (ASOs). (a) ASO_AL1, (b) ASO_AL2, (c) ASO_BR2, (d) ASO_BR1. PCR fragments are then sequenced to analyze the distribution of exchange points (**C**)

all these excess DSBs are repaired as NCOs but if so, the NCO number would be much higher than the number of COs. NCO tracts could be very short. Indeed, none of the NCO tracts reported so far in mice or human exceed 1.2 kb in length, with the average estimated at between 50 and 300 bp (1, 3, 5–7). Thus, if NCO tracts overlap only a few polymorphisms, detecting them will be more difficult.

In plants, only a few hotspots have been identified and characterized so far. All hotspots described have been discovered using phenotypic screens that allow detection of COs and sometimes NCOs. For example, the maize *bronze* locus hotspot was identified because a mutation in the corresponding gene changes the color of the kernel (8). Another well analyzed set of hotspots, *a1-sh2*, is flanked by two genes where mutations can change the color (*a1*) and composition (*sh2*) of maize kernels (9). In both regions, clusters of COs and NCOs have been found. However, differences compared to mammalian and yeasts hotspots have also been identified. At *bronze*, the NCO rate does not seem to exhibit a gradient with a peak in the center (10), although the CO distribution may not be sufficiently characterized to delimit the borders and, thus, the center. In the *a1-sh2* region, COs are detected over a large 10 kb region that could represent a cluster of overlapping hotspots (11) and the NCOs distribution may not have been properly analyzed because too few events are recovered.

To overcome the limitations of the phenotypic screens for characterization of plant meiotic hotspots, we used a molecular approach adapted from the "sperm-typing" technique thoroughly described in a series of recent papers (1, 2, 12). The procedure is based on sets of nested PCRs to isolate recombinant molecules from pools of parental nonrecombinant molecules (Fig. 2). We have previously described the method to characterize CO molecules (13). Here we report a modified protocol to extract high quality DNA from *Arabidopsis* pollen and two "pollen-typing"-based methods to identify, score and analyze NCO events (Fig. 3).

2 Materials

2.1 DNA Extraction

2.1.1 Extraction of DNA from Pollen

1. 10% Saccharose.

2. RNase A: 100 mg/ml.

3. DNeasy Plant Maxi Kit (Qiagen, Hilden, Germany).

4. TE buffer: 5 mM Tris–HCl (pH 8), 0.5 mM EDTA.

2.1.2 Extraction of DNA from Leaves

Use the established previously published protocol (13).

2.1.3 DNA Purification

Use the established previously published protocol (13).

2.2 Allele-Specific Long-Distance PCR

Use the established previously published protocol (13).

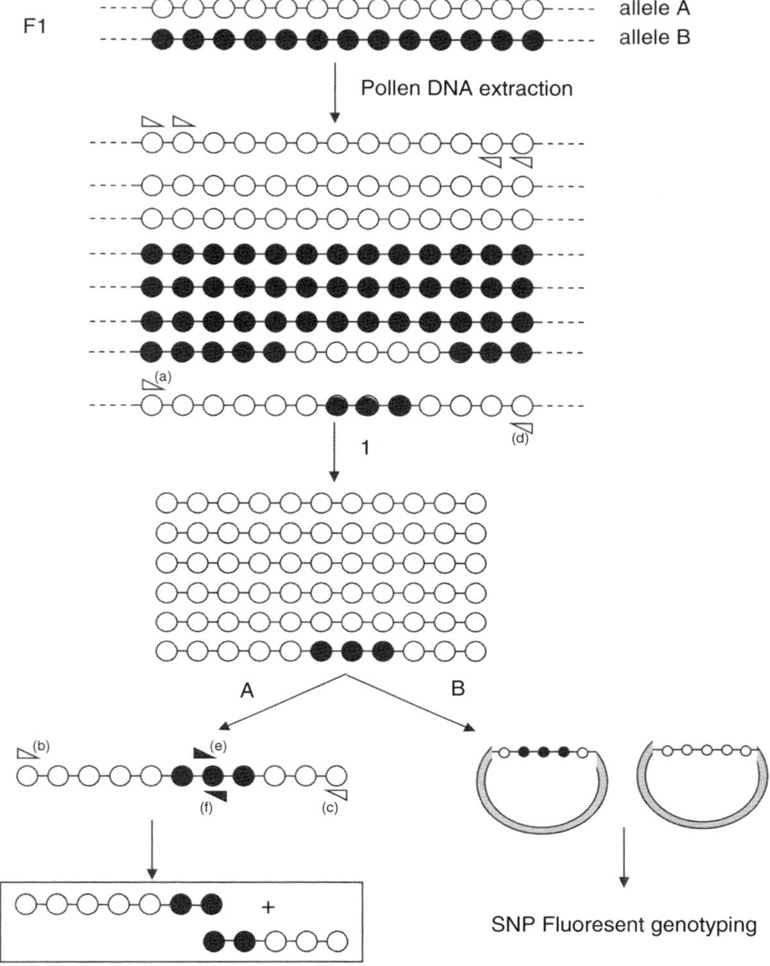

Fig. 3 Isolation of NCO events from pollen DNA. Genomic DNA is extracted from F1 (A × B) pollen. At a given hotspot locus, the majority of the molecules contain either the A or the B allele. A small fraction of the molecules contain NCO recombination products. Molecules containing flanking polymorphisms from the same parental allele are specifically amplified with one or two rounds of PCR with ASOs ASO_AL1 (a) and ASO_AR1 (d) (1). (A) NCOs are specifically amplified using two different PCRs conducted in parallel with two sets of ASOs. The fist one with primers ASO_AL2 (b) and ASO_BSnpLo (f) and the second one with primers ASO_AR2 (c) and ASO_BSnpUp (e). To score a NCO, both PCRs have to give positive results. Both PCR fragments are then sequenced to measure the length of the conversion event. (B) The central region containing the polymorphism to be tested for a NCO is cloned into a vector and individual clones are genotyped using fluorescently labeled oligonucleotides annealing to the polymorphic sites. DNA is extracted from positive clones and sequenced to confirm the NCO event and measure its length

3   Methods

3.1   *Genomic DNA Preparation*

In most experiments described below, especially those involving long-distance PCR amplification, the use of high molecular weight, high purity genomic DNA (gDNA) is required. This is not always a trivial matter, however, because mechanical disruption of the cell wall during DNA extraction often leads to extensive gDNA shearing, in particular in *Arabidopsis* (and most other plant species) pollen. Hence, careful optimization of the tissue lysis procedure is mandatory, keeping in mind that the optimal protocol should represent a compromise between yield (through high cell wall breakage efficiency) and integrity of gDNA.

Several methods for pollen disruption have been described, but few have proven to be adequate for high molecular weight gDNA isolation. Among them, grinding in liquid nitrogen, bead-vortexing and French pressing have been successfully used in our laboratory. We present here a protocol using a French press, which allows extracting a high yield of high molecular weight gDNA, provided that optimal pressure is used. The pressure value should be determined for each plant species prior to routine large-scale isolation of gDNA.

3.1.1 Extraction of Genomic DNA from Pollen

1. Harvest pollen from inflorescences as described in (13).

2. Resuspend the cell pellet in 5 ml of Buffer AP1 (Qiagen, Hilden, Germany).

   Lyse cells with a French press using pressure that will maximize the percentage of broken cells and minimizes shearing of DNA (see Note 1). Add 1 µl of the suspension to 10 µl of water onto a microscope glass side. Confirm the disruption of cells under a microscope. If lysis is successful, add RNase A to a final concentration of 0.2 mg/ml.

3. Proceed with DNA extraction using the Qiagen DNeasy Maxi Kit (Qiagen, Hilden, Germany) until the DNA elution step. Use three times 1 ml of TE Buffer (Qiagen, Hilden, Germany) for elution.

3.1.2 Extraction of Genomic DNA from Leaves

Use the established previously published protocol (13).

3.1.3 Purification of Genomic DNA

Use the established previously published protocol (13).

3.1.4 Quantification of Genomic DNA

Use the established previously published protocol (13).

3.2 Designing and Testing PCR Primers

Two kinds of oligonucleotides are used for PCR. These called universal oligonucleotides (UOs) anneal to identical sequences in both parental alleles. They should be designed following usual rules for PCR primer design (50% GC, avoid repeated regions, etc.). Their efficiency should be tested by using pairs of UOs and also in pairs with the allele-specific oligonucleotides (ASOs, see below).

ASOs are designed to anneal as specifically as possible only to one of the two parental alleles. Thus, their positions depend on the positions of sites polymorphic between the parental alleles. There are no clear rules to predict what will be a "highly" or "poorly" specific primer. The only requirement is that the 3′ terminal nucleotide of the ASOs must be a nucleotide polymorphic between the parental alleles. Thus, all ASOs have to be tested empirically (see below). To detect NCOs, ASOs are designed for polymorphism sites within the hotspot (Fig. 3). Additional criteria and methods to design and test the efficiency of UOs and ASOs, and the specificity of ASOs have been extensively described elsewhere (2, 12–14).

3.3 Detection and Characterization of Single NCO Molecules

Prior to any detection of recombinant molecules, genomic DNA extracted from pollen must be tested to determine its amplifiability (proportion of genomic molecules that can be PCR amplified). Long-distance PCRs up to 14 kb could be needed, depending on the width of the hotspot and the polymorphisms available at the hotspot borders. Typically, amplifiability ranges from 20 to 50% for amplicons longer than 8 kb and drops to around 10% for 14 kb. Thus, there could be a real discrepancy between the DNA concentration estimated by gel electrophoresis (see above) and amplifiability.

Methods to score amplifiable molecules are described in a previously published protocol (13).

NCO detection methods described below rely on the first allele-specific PCR to amplify only molecules with both flanking markers from the same allele (primers a and d in Fig. 3). If there is a suspicion that the first allele-specific PCR may not be highly specific, it is recommended to perform a second, nested allele-specific PCR. Then, on this amplified material, two methods can be used to identify NCOs (Fig. 3).

The simplest and quickest method is based on two allele-specific PCRs with two pairs of ASOs, performed in parallel. In each ASO pair, one ASO matches the flanking polymorphism from the allele amplified previously while the second ASO matches the polymorphism to be tested for the NCO and thus belonging to the other parental allele (A in Fig. 3). To identify a NCO, PCRs with both pairs of ASOs (primers b and f, and primers e and c) must give amplification products (Fig. 4).

If the polymorphisms do not allow designing specific ASOs, another method adapted from (15) can be used (B in Fig. 3). After two rounds of allele-specific PCRs, PCR fragments are digested and cloned into a vector. Individual clones are recovered in *E. coli* and tested for presence of NCOs by genotyping using fluorescently labeled oligonucleotides annealing to the polymorphic sites.

In this chapter, we use the following ASO designation:

–   ASO_AL1 (primer a in Fig. 3)/ASO_AR1 (primer d in Fig. 3), and ASO-AL2 (primer b in Fig. 3)/ASO_AR2 (primer c in Fig. 3) are two pairs of ASOs that specifically amplify allele A. The first pair is internal to the second pair.

–   ASO_BL1/ASO_BR1 and ASO-BL2/ASO_BR2 are two pairs of ASOs that amplify specifically allele B. The first pair is internal to the second pair.

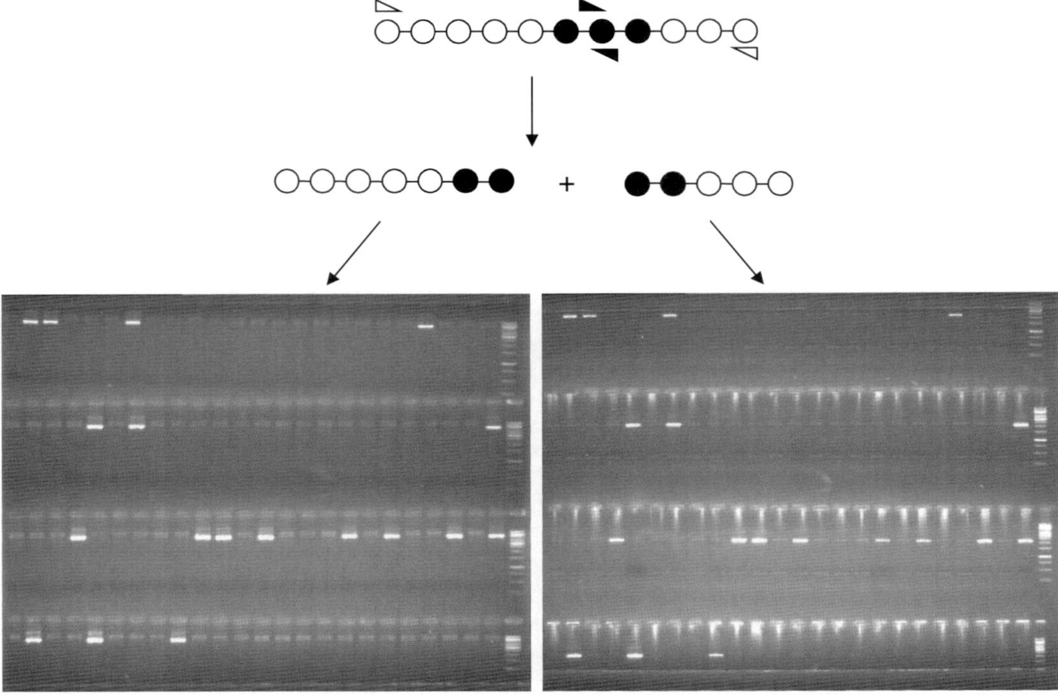

Fig. 4 Detection of SNPs using PCR. PCRs have been performed as described in Fig. 3A. Typical results are shown in the photos of the two gels. PCR products are present in corresponding lanes of the two gels, confirming positive NCO detection

– ASO_ASnpUp and ASO_ASnpLo are the upper and lower primers, respectively, specific to the SNP to be detected in allele A.

– ASO_BSnpUp (primer e in Fig. 3) and ASO_BSnpLo (primer f in Fig. 3) are the upper and lower primers specific to the SNP in allele B.

3.3.1   NCO Detection by PCR

The NCO activity of hotspots could be variable. NCO to CO ratios vary from 12:1 to 1:9 in mammals. Our own data (Laurène Giraut, unpublished) show NCO rates at a single site in *Arabidopsis* may vary between 1/100,000 and 1/250. Consequently, a series of different pollen DNA concentrations should be first tested to determine the ideal range before performing a large-scale analysis.

A. Optimizing gDNA Concentration

1. Prepare a premix for 14 reactions as follows:

   28 µl of 4 µM ASO_AL1

   28 µl of 4 µM ASO_AR1

   28 µl of 10× PCR buffer

   14 µl of 0.5 U/µl Taq:Pfu mix

   196 µl of H_2O

2. Combine 42 µl of the premix and 2 µl of 30 ng/µl F1 pollen gDNA in PCR tube "1."

3. Add 12 µl of H_2O to the remaining premix. Then, aliquot 22 µl of the premix to PCR tubes "2" to "12."

4. Transfer 22 µl from tube "1" to tube "2." Mix.

5. Transfer 22 µl from tube "2" to tube "3." Mix. Continue in this manner until tube "12." As a result of this procedure, gDNA is serially diluted by half in each consecutive tube, starting from 30 ng (roughly 200,000 *Arabidopsis* genomes).

6. Proceed to thermal cycling as follows:

   92°C for 2 min

   (92°C for 20 s, T_{opt} for 30 s, 68°C for 30 s + 45 s/kb) × 30

   68°C for 90 s/kb

   4°C

7. Proceed to secondary PCR to detect NCOs. Dilute 2 µl of the product of the first PCR in 100 µl of 5 mM Tris–HCl pH 8, 0.01% Triton X-100.

8. Then, prepare two reaction premixes for 14 reactions as follows:

   Premix 1:

   28 µl of 4 µM ASO_AL2

   28 µl of 4 µM ASO_BSnpLo

   28 µl of 10× PCR buffer

   14 µl of 0.5 U/µl Taq:Pfu mix

   168 µl of H_2O

   Premix 2:

   28 µl of 4 µM ASO_AR2

   28 µl of 4 µM ASO_BSnpUp

   28 µl of 10× PCR buffer

   14 µl of 0.5 U/µl Taq:Pfu mix

   168 µl of H_2O

9. Add 14 µl of the diluted PCR product obtained in the first PCR (see Subheading 3.3.1.A, step 7) to premix 1 and premix 2. Then, aliquot 22 µl to 12 tubes for each premix.

10. Proceed to thermal cycling as follows:

    92°C for 2 min

    (92°C for 20 s, T_{opt} for 30 s, 68°C for 45 s/kb) × 30

    68°C for 90 s/kb

    4°C

11. Add 5 µl of DNA loading dye to each PCR, and run 10 µl of the resulting mix in a 0.8% agarose gel (containing 0.2 µg/ml of ethidium bromide) in 1× TBE. Photograph the gel under UV. Use a long exposure time in order to see faint bands.

12. Find the lowest gDNA dilution that still results in PCR amplification and the highest gDNA dilution that is negative in both sets of second PCRs to perform a further series of 12 reactions to refine the optimal gDNA dilution. Choose the dilution that only results in one NCO product per reaction to avoid the risk of obtaining multiple NCO products in the same PCR (except if the NCO rate is low, see Note 2).

Perform the same analysis for the B allele. DNA strand bias during recombination initiation, leading to different NCO rates for each allele, has been described in several hotspots and in several species (16).

B. NCO detection

1. Prepare a reaction mix for 98 reactions as follows:

    196 µl of 4 µM ASO_AL1

    196 ml of 4 mM ASO_AR1

    196 ml of 10× PCR buffer

    98 µl of 0.5 U/µl Taq:Pfu mix

    69.3 µl of gDNA mixture containing 98× the optimal gDNA amount per PCR amount determined in Subheading 3.3.1.A

    126.7 ml of 5 ng/ml carrier DNA

    1274 µl of H_2O

2. Aliquot 22 µl of the mix into a PCR plate

3. Proceed to thermal cycling as follows:

    92°C for 2 min

    (92°C for 20 s, T_{opt} for 30 s, 68°C for 45 s/kb) × 30

    68°C for 90 s/kb

    4°C

4. Proceed to secondary PCR. Dilute 2 µl of the products of the first PCR in 100 µl of 5 mM Tris–HCl pH 8, 0.01% Triton X-100.

5. Prepare two reaction premixes for 98 reactions as follows:

    Premix 1:

    196 ml of 4 mM ASO_AL2

    196 ml of 4 mM ASO_BSnpLo

    96 ml of 10× PCR buffer

    98 ml of 0.5 U/ml Taq:Pfu mix

    1372 µl of H_2O

    Aliquot 21 ml of premix 1 into PCR plate 2a

Premix 2:

196 ml of 4 mM ASO_AR2

96 ml of 4 mM ASO_BSnpUp

196 ml of 10× PCR buffer

98 ml of 0.5 U/ml Taq:Pfu mix

1372 ml of H2O

Aliquot 21 µl of premix 2 into the wells of PCR plate 2b

6. Aliquot 1 µl of diluted product from the first PCR to PCR plates 2a and 2b.

Proceed to thermal cycling as follows:

92°C for 2 min

(92°C for 20 s, Topt for 30 s, 68°C for 45 s/kb) ×30

68°C for 90 s/kb

4°C

7. Add 5 µl of DNA loading dye to each PCR, and run 10 µl of the mixture in a 0.8% agarose gel (containing 0.2 µg/ml of ethidium bromide) in 1× TBE. Photograph the gel under UV. Use a long exposure time to see faint bands.

8. An NCO is detected when both sets of secondary PCRs (performed with ASO_AL2/ASO_BSnpLo and ASO_AR2/ASO_BSnpUp) are positive for the same PCR product obtained in the first PCR (Fig. 4). The NCO rate is the ratio between the number of clear positive reactions and the number of input genomes in PCR1. NCO tract lengths are determined by sequencing the PCR products.

3.3.2 NCO Detection by Cloning and High Throughput Genotyping

1. Allele-specific PCRs are first carried out on pollen genomic DNAs. Typically, a series of 12–24 reactions, each containing from 1 to 10,000 amplifiable molecules, are conducted with pairs of ASOs specific to each of the parental alleles (Fig. 3, see Subheading 3.3.1). Thus, after amplification, each set of reactions should contain a majority of nonrecombinant PCR products of one parental allele and along with some molecules with NCOs. In parallel, amplify the same region from genomic DNAs extracted from leaves of each parent. It will be used as a control in the fluorescent assay.

2. Then, analyze PCR products by gel electrophoresis. If amplifications are of good quality (one unique strong band of the correct size), pool PCR products with the same pair of ASOs. If several bands are seen in gel electrophoresis after the first allelic PCR, perform a secondary, nested PCR with ASOs specific to the same parental allele and control the quality of the amplification.

3. Clone the region containing the polymorphisms to be tested for NCOs in an appropriate cloning vector (see Note 3).

In parallel, clone the parental allele amplified in the PCR using leaf DNA.

4. Grow *E. coli* colonies in 96 wells plates at 37°C for 18 h. Centrifuge cultures and resuspend the cell pellets in 150 µl water. There is no need to extract plasmid DNA, genotyping is performed directly on cell cultures.

5. Genotype clones using the Chemicon Amplifluor SNPs Genotyping System (Millipore, Billerica, MA, USA). For this purpose, oligonucleotides specific to SNPs in each parental allele are designed using the Amplifluor AssayArchitect software (https://apps.serologicals.com/AAA/). Perform SNP genotyping as recommended by the manufacturer. Appropriate controls should be included. In each plate, at least 2 wells with H_2O only and 2 wells with cloned fragments of each parental allele should be used. To increase the number of colonies genotyped, pool two or three colonies in the same well after growing the cell cultures. In this case, positive colonies will give a "heterozygous" signal (Fig. 5). The same plate can be used to genotype several different polymorphisms independently.

6. Extract DNA from "positive" colonies to confirm the NCO event by DNA sequencing and determine the length of the meiotic event.

4 Notes

1. For the pollen suspension prepared from the Columbia × Landsberg *erecta* F1 we used the pressure of 3,000 psi. However, we recommend testing different levels of pressure for each cross. After lysis, the percentage of broken cells should be evaluated and the average DNA size should be estimated by gel electrophoresis. Then, the pressure that maximizes the percentage of broken pollen cells and minimizes DNA fragmentation should be selected for DNA extraction on the rest of the pollen suspension.

2. If the NCO rate is low, it is better to multiply reactions with fewer genomes than to carry out fewer reactions with high numbers of genomes. PCR may result in nonspecific amplification products when nonrecombinant molecules are in large excess. The tolerated excess depends on the quality of ASOs and usually is between a 1,000 and 10,000.

3. It is important to optimize the cloning efficiency by selecting appropriate restriction enzyme(s), dephosphorylating vector ends before ligation, and/or using additional restriction enzymes to avoid self-circularization of the cloning vector. The efficiency of cloning should be estimated before downstream

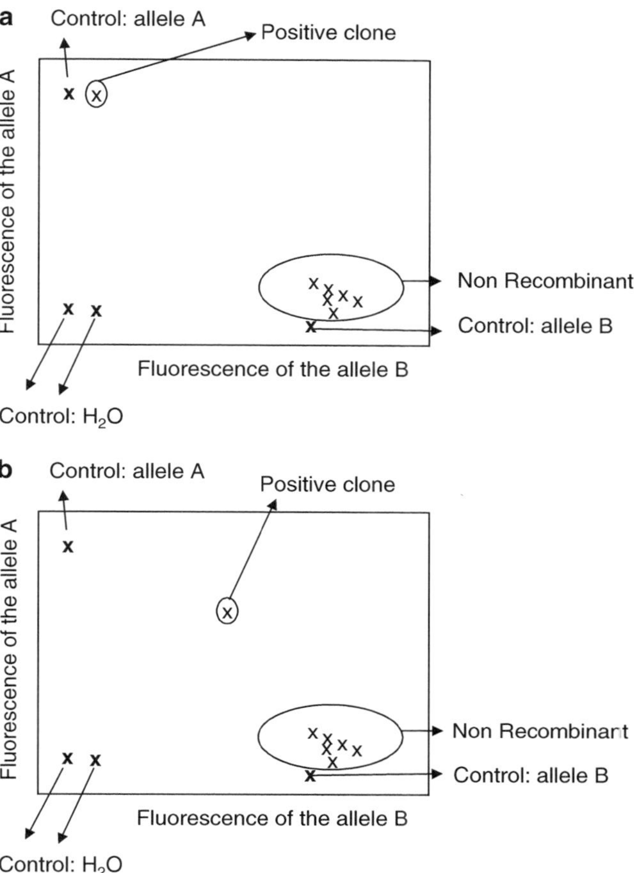

Fig. 5 Detection of NCOs using a fluorescent assay. (**a**) Fluorescent assay on individual colonies. Controls with only water do not exhibit any fluorescence. Parental control clones exhibit specific fluorescence. Clones containing nonrecombinant DNA exhibit the same fluorescence as the parental B allele whereas positive clones exhibit fluorescence of the A allele. (**b**) Fluorescent assay on pooled clones. Controls are the same as above. After cell growth, clones are pooled by two or three. Pools that do not contain any recombinant molecules exhibit the same fluorescence as the parental B allele whereas pools containing one recombinant molecule exhibit fluorescence for both A and B alleles

analyses either with the blue/white screening system or by PCR on *E. coli* colonies with adapter primers. Ideally, each colony should represent a unique molecule cloned from pollen genomic DNA. The number of haploid genomes to be used in the first specific PCR depends on the number of colonies that will be tested. The ratio of the number of colonies to the number of amplifiable haploid genomes will give the probability (m) that a given pollen genomic DNA molecule is represented among the selected clones. Ng et al. (15) have estimated the probability of any 2 (n) clones being duplicates of the same initial event using a Poisson estimation $P(n) = (e^{-m} m^{n}) \, n!$. In the example they provide, 14 meiotic events were detected in 500 clones obtained from an input of 11,000 haploid genomes. These numbers give a 1.25% probability that the 14 clones are duplicates of the same initial molecule.

Acknowledgments

We thank the members of the "Méiose et Recombinaison" group for helpful discussions. Research leading to this protocol was funded by the European Community Seventh Framework Programme (FP7/2007-2013 FP7/2007-2011) under grant agreement KBBE-2009-222883.

References

1. Baudat F, de Massy B (2007) Regulating double-stranded DNA break repair towards crossover or non-crossover during mammalian meiosis. Chromosome Res 15:565–577

2. Kauppi L, May CA, Jeffreys AJ (2009) Analysis of meiotic recombination products from human sperm. Methods Mol Biol 557:323–355

3. Arnheim N, Calabrese P, Tiemann-Boege I (2007) Mammalian meiotic recombination hot spots. Annu Rev Genet 41:369–399

4. de Muyt AD, Mercier R, Mezard C, Grelon M (2009) Meiotic recombination and crossovers in plants. Genome Dyn 5:14–25

5. Jeffreys AJ, Neumann R (2005) Factors influencing recombination frequency and distribution in a human meiotic crossover hotspot. Hum Mol Genet 14:2277–2287

6. Jeffreys AJ, Holloway JK, Kauppi L, May CA, Neumann R, Slingsby MT et al (2004) Meiotic recombination hot spots and human DNA diversity. Philos Trans R Soc Lond B Biol Sci 359:141–152

7. Holloway K, Lawson VE, Jeffreys AJ (2006) Allelic recombination and de novo deletions in sperm in the human beta-globin gene region. Hum Mol Genet 15:1099–1111

8. Dooner HK (1986) Genetic fine structure of the *BRONZE* locus in maize. Genetics 113:1021–1036

9. Xu X, Hsia AP, Zhang L, Nikolau BJ, Schnable PS (1995) Meiotic recombination break points resolve at high rates at the 5′ end of a maize coding sequence. Plant Cell 7:2151–2161

10. Dooner HK, Martinez-Ferez IM (1997) Recombination occurs uniformly within the bronze gene, a meiotic recombination hotspot in the maize genome. Plant Cell 9:1633–1646

11. Yao H, Zhou Q, Li J, Smith H, Yandeau M, Nikolau BJ et al (2002) Molecular characterization of meiotic recombination across the 140-kb multigenic *a1-sh2* interval of maize. Proc Natl Acad Sci USA 99:6157–6162

12. Cole F, Jasin M (2011) Isolation of meiotic recombinants from mouse sperm. Methods Mol Biol 745:251–282

13. Drouaud J, Mezard C (2011) Characterization of meiotic crossovers in pollen from *Arabidopsis thaliana*. Methods Mol Biol 745:223–249

14. Baudat F, de Massy B (2009) Parallel detection of crossovers and noncrossovers in mouse germ cells. Methods Mol Biol 557:305–322

15. Ng SH, Parvanov E, Petkov PM, Paigen K (2008) A quantitative assay for crossover and noncrossover molecular events at individual recombination hotspots in both male and female gametes. Genomics 92:204–209

16. Baudat F, de Massy B (2007) Cis- and trans-acting elements regulate the mouse Psmb9 meiotic recombination hotspot. PLoS Genet 3:e100

Chapter 19

Chromatin Immunoprecipitation for Studying Chromosomal Localization of Meiotic Proteins in Maize

Yan He, Gaganpreet Sidhu, and Wojciech P. Pawlowski

Abstract

Chromatin immunoprecipitation (ChIP) is a method that allows identification of chromosomal sites occupied by specific proteins. In this technique, chromatin is extracted from cells, sheared, and, using a specific antibody, enriched in fragments that contain a protein of interest. Genomic location of the protein can then be identified by hybridization of the resulting DNA to tiling microarrays or by sequencing. Thanks to advances in high-throughput sequencing methods, studying protein localization using ChIP has become possible even in species with relatively large genomes. Here, we describe a ChIP protocol that we developed to examine localization of meiotic proteins in maize.

Keywords Chromosomes, Chromatin, Immunoprecipitation, Antibody, Maize, Recombination

1 Introduction

Many proteins involved in meiosis form complexes that are visible as distinct foci on chromosomes during prophase I. Some of these proteins are components of the recombination pathway and mark sites of DNA double-strand breaks (DSBs), recombination pathway intermediates, such as single-end invasion (SEI) events, and crossovers (COs). Key recombination proteins, such as SPO11, RAD51 (Fig. 1), DMC1, and MLH1 (1–6), belong to this group. Specifically modified chromatin proteins that are implicated in meiotic prophase progression also occur at discrete chromosomal locations. These modifications include histone H3K4 tri-methylation, thought to mark sites of DSB hotspots in some species, and γ-H2AX phosphorylation, which marks chromatin regions adjacent to nascent DSBs (6–8). Finally, synaptonemal complex proteins, such as ZYP1, often show localization at distinct sites on chromosomes during early stages of their installation (5).

Protein distribution on meiotic chromosomes is often examined using immunolocalization microscopy (see Chapters 9–11).

Wojciech P. Pawlowski et al. (eds.), *Plant Meiosis: Methods and Protocols*, Methods in Molecular Biology, vol. 990, DOI 10.1007/978-1-62703-333-6_19, © Springer Science+Business Media New York 2013

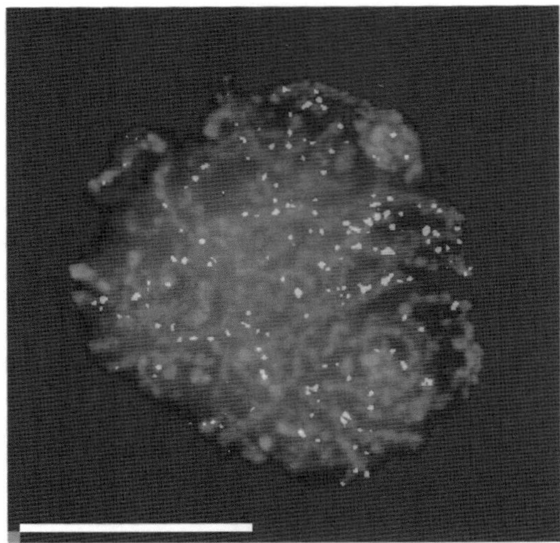

Fig. 1 The RAD51 protein localized to discrete sites on maize chromosomes during zygotene. Chromosome sites where RAD51 is located can be determined with high resolution using chromatin immunoprecipitation (ChIP). *Red* = chromatin. *Green* = RAD51. Bar = 10 μm. Modified from Pawlowski et al. (3) copyright American Society of Plant Biologists (www.plantcell.org)

Another approach for studying patterns of protein distribution along chromosomes is chromatin immunoprecipitation (ChIP) (9, 10). In this method, chromosomal proteins are cross-linked and chromatin is extracted and enriched, using a specific antibody, in fragments containing the protein of interest. The antibody-enriched fragments can be identified using either whole-genome DNA tiling arrays or next-generation sequencing.

The ChIP technique offers several advantages over microscopical immunolocalization studies. First, chromosomal locations of the target protein can be determined at a very high resolution of a few hundred base pairs. Second, a very large number of meiotic cells, typically millions of them, are surveyed in a single experiment. Finally, relative frequencies of target protein occupancy can be determined for specific chromosome locations. These features can provide deep insights into how meiotic processes operate on a genome-wide scale. Good examples of how using the ChIP technique can further understanding of meiosis are studies on mapping the sites of meiotic DSBs in yeast and mouse using immunoprecipitation of chromatin fragments associated with the SPO11 and RAD51 proteins (11, 12). SPO11, a protein belonging to the topoisomerase family (13), is responsible for generating DSBs in chromosomal DNA that initiate meiotic recombination. The DSBs are subsequently resected to form single-stranded DNA ends that become coated by a complex containing two recombination proteins, RAD51 and DMC1 (14–16). Consequently, SPO11-associated chromosome fragments represent the sites of

DSBs while fragments associated with RAD51 are immediately adjacent to the DSB sites.

A key requirement for ChIP experiments is availability of antibodies with high specificity and affinity to native proteins. Performance of an antibody in immunolocalization and/or western blot experiments may not always be a sufficient predictor of its suitability for ChIP. Actual ChIP experiments, with appropriate negative controls, need to be carried out to determine antibody's performance. While a large number of antibodies against meiotic proteins are commercially available for mouse and humans, many of them do not show high enough affinity to plant proteins to be useful for ChIP. Consequently, many ChIP experiments in plant may require rising custom antibodies.

In this chapter, we describe a ChIP protocol to study distribution of chromosomal proteins in male meiosis in maize. We developed and optimized this protocol using an antibody against H3K4 tri-methylation and we utilize it to determine locations of meiotic DSB sites using an antibody against RAD51.

2   Materials

2.1   Plants

Starting with high-quality plant material is critical for success of ChIP experiments. Since temperature is known to affect the progression of meiosis (17), grow plants in a controlled-environment growth chamber. To grow maize plants, use a 12 h day/12 h night photoperiod, temperature of 31°C during the day and 22°C at night, and light intensity of about 600 $\mu mol/m^2/s$.

2.2   Reagents

2.2.1   Staging Meiotic Flowers

1. Farmer's fixative: 3 volumes of 100% ethanol, 1 volume of glacial acetic acid.

2. Acetocarmine: 2% acetocarmine powder in 45% acetic acid. Boil for 6–8 h in a flask with an attached reflux column and then filter through filter paper when the solution is still warm. Store in a dark bottle at room temperature.

2.2.2   Cross-linking and Chromatin Preparation

1. Cross-linking buffer: 10 mM Tris–HCl (pH 8.0), 0.4 M sucrose, 10 mM $MgCl_2$, 5 mM β-mercaptoethanol, 1% formaldehyde.

2. 2 M glycine in water.

3. Distilled water.

4. Chromatin extraction buffer A: 10 mM Tris–HCl (pH 8.0), 0.4 M sucrose, 10 mM $MgCl_2$, 1 mM phenylmethylsulfonyl fluoride (PMSF), 5 mM β-mercaptoethanol. Before use, add 1 tablet of Complete Protease Inhibitor (Roche Applied Science, Indianapolis, IN, USA) per 50 ml of buffer.

5. Chromatin extraction buffer B: 10 mM Tris–HCl (pH 8.0), 0.25 M sucrose, 10 mM MgCl$_2$, 1% Triton X-100, 1 mM phenylmethylsulfonyl fluoride (PMSF), 5 mM β-mercaptoethanol, 1 μg/ml Protease Inhibitor Cocktail (Sigma Aldrich, St. Louis, MO, USA).

6. Chromatin extraction buffer C: 10 mM Tris–HCl (pH 8.0), 1.7 M sucrose, 2 mM MgCl$_2$, 0.15% Triton X-100, 1 mM phenylmethylsulfonyl fluoride (PMSF), 5 mM β-mercaptoethanol, 1 μg/ml Protease Inhibitor Cocktail (Sigma Aldrich, St. Louis, MO, USA).

7. Nuclei lysis buffer: 50 mM Tris–HCl (pH 8.0), 10 mM EDTA, 1% (w/v) SDS, 1 mM phenylmethylsulfonyl fluoride (PMSF), 1 μg/ml Protease Inhibitor Cocktail (Sigma Aldrich, St. Louis, MO, USA).

2.2.3 Chromatin Immunoprecipitation

1. ChIP dilution buffer: 16.7 mM Tris–HCl (pH 8.0), 1.2 mM EDTA, 167 mM NaCl, 1.1% Triton X-100, 1 mM phenylmethylsulfonyl fluoride (PMSF), 1 μg/ml Protease Inhibitor Cocktail (Sigma Aldrich, St. Louis, MO, USA).

2. Blocking buffer: 16.7 mM Tris–HCl (pH 8.0), 1.2 mM EDTA, 167 mM NaCl, 1.1% Triton X-100, 1 mM phenylmethylsulfonyl fluoride (PMSF), 1 μg/ml Protease Inhibitor Cocktail (Sigma Aldrich, St. Louis, MO, USA), 5 mg/ml bovine serum albumin (BSA).

3. Low salt wash buffer: 20 mM Tris–HCl (pH 8.0), 2 mM EDTA, 150 mM NaCl, 0.1% SDS, 1% Triton X-100.

4. High salt wash buffer: 20 mM Tris–HCl (pH 8.0), 2 mM EDTA, 500 mM NaCl, 0.1% SDS, 1% Triton X-100.

5. LiCl wash buffer: 10 mM Tris–HCl (pH 8.0), 1 mM EDTA, 250 mM LiCl, 1% NP-40, 1% sodium deoxycholate.

6. TE buffer: 10 mM Tris–HCl (pH 8.0), 1 mM EDTA.

7. Elution buffer: 50 mM Tris–HCl (pH 8.0), 10 mM EDTA, 200 mM NaCl, 1% SDS.

8. 10 mg/ml RNase.

9. 20 mg/ml Proteinase K.

10. MinElute PCR Purification Kit (Qiagen, Hilden, Germany).

11. Quant-IT dsDNA HS Assay Kit (Invitrogen, Grand Island, NY).

2.3 Supplies and Equipment

2.3.1 Staging and Collecting Meiotic Flowers

1. Glass scintillation vials or 15 ml plastic tubes to collect flowers for staging.

2. Razor blade.

3. Tweezers with fine tips.

4. Dissecting needle.

5. Rusty nail (see Note 1).

6. Glass microscope slides and coverslips.

7. Alcohol lamp.

8. Masking tape.

9. Plastic tray with wet paper towels.

10. Dissecting stereoscope.

11. Bright-field microscope.

2.3.2 Cross-linking and Chromatin Preparation

1. 50 ml conical tubes.

2. Miracloth.

3. Small ceramic mortar and pestle.

4. Plastic funnel.

5. Vacuum desiccator

6. Probe sonicator.

7. Microcentrifuge.

8. Refrigerated centrifuge.

9. Tabletop shaker.

10. 4°C refrigerator.

2.3.3 Chromatin Immunoprecipitation

1. Dynabeads (Invitrogen, Grand Island, NY).

2. Magnetic Separation Stand (Invitrogen, Grand Island, NY).

3. Tube rotator.

3 Methods

3.1 Staging Meiotic Flowers

1. At the time of meiosis, the immature tassel in maize is still inside the stalk. The presence of the tassel can be felt just below the top node of the plant by gently squeezing the leaf whorl.

2. After establishing that the tassel is large enough to be felt, make a small longitudinal incision with a razor blade through the leaves to the tassel, just below the top node.

3. Remove several flowers with needle-nosed forceps into a glass scintillation vial or tube containing Farmer's fixative.

4. Dissect anthers from the collected flowers on a microscope slide under a stereo dissecting microscope.

5. Add a drop of acetocarmine solution for staining. Mix anthers with the stain using a dissecting needle or, preferably, a rusty nail (see Note 1) over gentle heat until the color of the stain turns from deep red to purple. Place a coverslip over the anthers and gently press on to break the anthers and release meiocytes. Determine the stage of meiosis under a bright-field compound microscope.

6. If the anthers are not yet at the desired meiosis stage, tape over the incision with masking tape and repeat the staging procedure in a day or two.

7. Cut plants with anthers at the desired stage of meiosis at several nodes below the tassel.

8. Gently remove leaves surrounding the tassel. To prevent the tassel from drying out during dissection, place it on wet paper towels in a tray and put more wet paper towels on top of it.

3.2 Chromatin Cross-linking

1. Harvest 1.5 g of healthy-looking flower tissue (see Note 2) and transfer it into a 50 ml conical tube.

2. Gently submerge the tissue in 37 ml of cross-linking buffer (see Note 3). Cap the tube with Miracloth to prevent the tissue from floating on the surface.

3. Vacuum infiltrate for 10 min.

4. Release vacuum slowly and remove Miracloth. Stop the cross-linking reaction by adding 2.5 ml of 2 M glycine. Vacuum infiltrate for 5 min.

5. Decant supernatant and wash the tissue three times with 40 ml of distilled water. After the third wash, remove as much water as possible by putting the tissue between dry paper towels.

6. Transfer dry tissue into a new 50 ml conical tube. Snap-freeze in liquid nitrogen and store at −80°C.

3.3 Chromatin Preparation

1. Grind the tissue to a fine power with a precooled mortar and pestle.

2. Resuspend the powder in 40 ml of ice-cold chromatin extraction buffer A. Incubate for 20 min at 4°C with gentle shaking.

3. Filter solution into a new 50 ml conical tube through 2 layers of Miracloth placed in a plastic funnel.

4. Centrifuge at $1,250 \times g$ for 20 min at 4°C.

5. Pour out the supernatant and resuspend the pellet in 1 ml of ice-cold extraction buffer B by gently pipetting up and down with a 1,000 µl automatic pipette. Transfer the suspension to a 1.5 ml microcentrifuge tube. Incubate on ice for 15 min with occasional agitation.

6. Centrifuge at $18,500 \times g$ in a microcentrifuge for 10 min at 4°C. Discard the supernatant and resuspend the pellet in 500 µl of ice-cold extraction buffer C by gently pipetting up and down with a 1,000 µl automatic pipette (see Note 4).

7. In a clean 1.5 ml microcentrifuge tube, add 500 µl of extraction buffer C. Layer the resuspended pellet from step 6 on top of this "cushion."

8. Centrifuge at 14,000 rpm in a microcentrifuge for 1 h at 4°C.

9. Discard the supernatant and resuspend the nuclei pellet in 500 µl of ice-cold nuclei lysis buffer.

3.4 Chromatin Sonication

1. Sonicate chromatin on ice using several pulses lasting for 5 s each to shear DNA into fragments of desirable size (see Notes 5 and 6).

2. After sonication, centrifuge the chromatin solution at 14,000 rpm in a microcentrifuge for 5 min at 4°C to pellet tissue debris. Transfer the supernatant to a new tube.

3. Take a 10 µl aliquot of the sonicated chromatin sample to check sonication efficiency and use as an input control sample for ChIP product sequencing (see Note 7).

3.5 Blocking Protein A Beads (see Notes 8 and 9)

1. For each ChIP sample, take 100 µl of Dynabeads (Invitrogen, Grand Island, NY) slurry into a 1.5 ml microcentrifuge tube.

2. Separate the beads on a Magnetic Separation Stand (Invitrogen, Grand Island, NY) for 1 min. Without disturbing the beads, pipette out the supernatant.

3. Wash beads twice with 1 ml of ChIP dilution buffer. For each wash, add the buffer and vortex the beads briefly to break coagulates. Then, remove the buffer as described in step 2.

4. Resuspend the beads in 1 ml of blocking buffer. Incubate at 4°C with gentle shaking for at least 2 h.

5. Wash the beads three times with 1 ml of ChIP dilution buffer as described in step 3.

6. Add ChIP dilution buffer back to the original bead volume from step 1.

3.6 Chromatin Immunoprecipitation

1. Split the chromatin sample from Subheading 3.4, step 3 (approx. 450 µl) into three 1.5 ml tubes of equal volume (150 µl in each tube) and dilute the chromatin sample in each tube tenfold by adding 1,350 µl of ChIP dilution buffer (see Note 10).

2. Preclear each chromatin sample by mixing with 40 µl of protein A-coated beads for 3 h with gentle rotation of the tubes on a tube rotator at 4°C.

3. Separate the beads on the Magnetic Separation Stand.

4. Transfer the supernatant from each tube into a new tube. The first tube will serve as a "no-antibody" control. Add 10 µg of normal rabbit IgG to the second tube to use as an "IgG control." Add 10 µg of your target antibody to the third tube (see Note 11).

5. Incubate the chromatin samples overnight with rotating at 4°C.

6. The next morning, capture the protein–DNA complexes by adding 40 µl of coated-beads and rotating the tubes on a tube rotator for 2.5 h at 4°C. Separate beads on the Magnetic Separation Step and remove the supernatant.

7. Wash the beads with 1 ml of each of low salt wash buffer, high salt wash buffer, LiCl wash buffer, and TE by rotating the tubes for 5 min at 4°C. Conduct each wash as described in Subheading 3.5, step 2. After the final wash, make sure to remove all TE.

3.7 Eluting Immunoprecipitated Complexes (See Note 12)

1. Add 200 µl of freshly prepared elution buffer. Resuspend beads by vortexing and incubate at 65°C for 30 min with occasional agitation.

2. Centrifuge at $2,000 \times g$ for 1 min and collect the supernatant into a new tube.

3.8 De-Cross-linking

1. Add 4 µl of 10 mg/ml RNase to the supernatants from Subheading 3.7, step 2, and incubate at 37°C for 1.5 h.

2. Add 4 µl of 20 mg/ml Proteinase K and incubate at 45°C for 2 h.

3. Reverse cross-link at 65°C for 8 h or overnight.

3.9 DNA Recovery

1. Purify DNA from each sample Subheading 3.8, step 3 using the MinElute PCR Purification Kit (Qiagen, Hilden, Germany), eluting in 30 µl of H_2O.

2. Measure the DNA concentration with the Quant-IT dsDNA HS assay kit (Invitrogen, Grand Island, NY) following manufacturer's instructions.

4   Notes

1. Iron oxide that leaches from the rusty nail enhances the staining reaction. Too little iron oxide will result in weaker staining.

2. This protocol is primarily designed for immunoprecipitating proteins that are specific to meiosis and are not present, or not associated with chromatin, in somatic cells. Targeting proteins that accumulate on chromatin in somatic tissues will lead to confounded results. In such cases, isolated meiocytes (see Chapter 20) can be used instead of whole flowers. This modification will require a scaled-down ChIP protocol to accommodate a much smaller amount of input material.

3. The length of cross-linking and the formaldehyde concentration need to be optimized for each type of tissue. Insufficient cross-linking may result in decreased binding of the ChIP antibody, while excessive cross-linking may lead to nonspecific binding of the antibody.

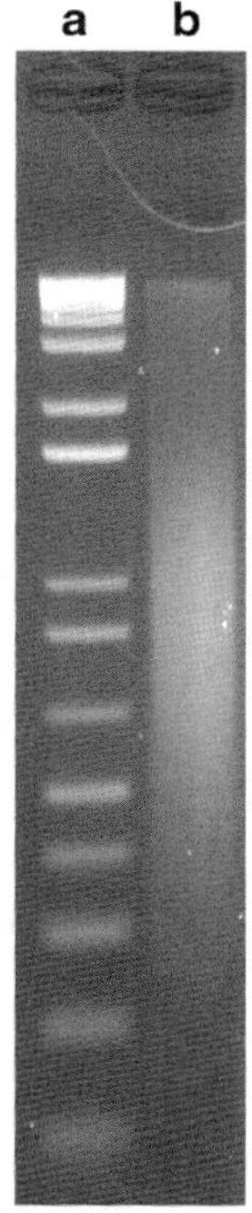

Fig. 2 Analysis of chromatin sonication efficiency. (**a**) 200 ng of sonicated chromatin de-cross-linked and run in a 2% agarose gel. (**b**) 1 kb DNA molecular size ladder (Invitrogen, Grand Island, NY)

4. Avoid introducing air bubbles or forming froth on the surface as this may lead to protein degradation in subsequent steps.

5. In most experiments, chromatin is fragmented to sizes of 200–500 bp. However, in our RAD51 ChIP, we shear DNA into 500–1,000 bp fragments (using 8 sonicator pulses of 5 s each). To produce desired fragment sizes, sonication conditions need to be optimized for every sonicator type, type of tissue, and, in some cases, also for each specific target protein. Over-sonication will lead to DNA degradation, whereas insufficient sonication will lead to nonspecific antibody binding and decreased ChIP yield. See also Note 7.

6. Keep sample on ice to avoid generating excessive heat during sonication, since it will cause protein degradation. Leave samples on ice for at least 30 s between each sonicator pulse to allow them to cool down.

7. The sonicated chromatin sample needs to be de-cross-linked before DNA extraction. To do this, add 140 μl of TE buffer, 5 μl of 5 M NaCl, and 10 μl of 10% SDS to a 10 μl aliquot of sonicated chromatin. Reverse-cross-link overnight at 65°C. Purify DNA using the Qiagen MinElute PCR Purification Kit (Qiagen, Hilden, Germany). To check sonication efficiency, electrophorese an aliquot of the extracted DNA in 2% agarose gel (Fig. 2).

8. Blocking the beads decreases nonspecific binding of the antibody. We strongly recommend including this step, even though it is not always suggested in published ChIP protocols. Do not use DNA as a blocking reagent if the ChIP DNA product will be analyzed by sequencing. Otherwise, most of the sequence reads will represent carrier DNA.

9. The bead blocking step can be carried out before starting the ChIP experiment. After blocking, beads can be stored at 4°C.

10. From this step on, use low-retention microcentrifuge tubes.

11. The amount of antibody used may need to be optimized for each antibody. This amount is dependent on the affinity between the antibody and the antigen that varies from one antibody to another.

12. In some cases, a second immunoprecipitation may be desired to increase specificity. In this case, chromatin elution after the first immunoprecipitation round should be performed at room temperature so that protein–DNA complexes are not denatured. The second immunoprecipitation round should be performed in the same way as the first immunoprecipitation, except the elution step, which should be carried out at 65°C.

Acknowledgment

Research to develop this protocol was supported by a grant from National Science Foundation (IOS-1025881) to W.P.P.

References

1. Terasawa M, Shinohhara A, Hotta Y, Ogawa H, Ogawa T (1995) Localization of RecA-like protein in chromosomes of the lily at various meiotic stages. Genes Dev 9:925–934

2. Franklin AE, McElver J, Sunjevaric I, Rothstein R, Bowen B, Cande WZ (1999) Three-dimensional microscopy of the Rad51 recombination protein during meiotic prophase. Plant Cell 11:809–824

3. Pawlowski WP, Golubovskaya IN, Cande WZ (2003) Altered nuclear distribution of recombination protein RAD51 in maize mutants suggests the involvement of RAD51 in meiotic homology recognition. Plant Cell 15:1807–1816

4. Chelysheva L, Grandont L, Vrielynck N, le Guin S, Mercier R, Grelon M (2010) An easy protocol for studying chromatin and recombination protein dynamics during *Arabidopsis thaliana* meiosis: immunodetection of cohesins, histones and MLH1. Cytogenet Genome Res 129:143–153

5. Higgins JD, Sanchez-Moran E, Armstrong SJ, Jones GH, Franklin FCH (2005) The Arabidopsis synaptonemal complex protein ZYP1 is required for chromosome synapsis and normal fidelity of crossing over. Genes Dev 19:2488–2500

6. Sanchez-Moran E, Santos JL, Jones GH, Franklin FCH (2007) ASY1 mediates AtDMC1-dependent interhomolog recombination during meiosis in *Arabidopsis*. Genes Dev 21:2220–2233

7. Borde V, Robine N, Lin W, Bonfils S, Geli V, Nicolas A (2009) Histone H3 lysine 4 trimethylation marks meiotic recombination initiation sites. EMBO J 28:99–111

8. Buard J, Barthes P, Grey C, de Massy B (2009) Distinct histone modifications define initiation

and repair of meiotic recombination in the mouse. EMBO J 28:2616–2624

9. O'Neill LP, Turner BM (1996) Immunoprecipitation of chromatin. Methods Enzymol 274:189–197

10. Collas P (2010) The current state of chromatin immunoprecipitation. Mol Biotechnol 45: 87–100

11. Smagulova F, Gregoretti IV, Brick K, Khil P, Camerini-Otero RD, Petukhova GV (2011) Genome-wide analysis reveals novel molecular features of mouse recombination hotspots. Nature 472:375–378

12. Prieler S, Penkner A, Borde V, Klein F (2005) The control of Spo11's interaction with meiotic recombination hotspots. Genes Dev 19:255–269

13. Keeney S, Giroux CN, Kleckner N (1997) Meiosis-specific DNA double-strand breaks are catalyzed by Spo11, a member of a widely conserved protein family. Cell 88:375–384

14. Bishop DK, Park D, Xu L, Kleckner N (1992) DMC1: A meiosis-specific yeast homolog of *Escherichia coli* recA required for recombination, synaptonemal complex formation, and cell cycle progression. Cell 69:439–456

15. Masson J-Y, West SC (2001) The Rad51 and Dmc1 recombinases: a non-identical twin relationship. Trends Bioch Sci 26:131–136

16. Shibata T, Nishinaka T, Mikawa T, Aihara H, Kurumizaka H, Yokoyama S et al (2001) Homologous genetic recombination as an intrinsic dynamic property of a DNA structure induced by RecA/Rad51-family proteins: a possible advantage of DNA over RNA as genomic material. Proc Natl Acad Sci USA 98:8425–8432

17. Francis KE, Lam SY, Harrison BD, Bey AL, Berchowitz LE, Copenhaver GP (2007) Pollen tetrad-based visual assay for meiotic recombination in *Arabidopsis*. Proc Natl Acad Sci USA 104:3913–3918

Chapter 20

Analyzing the Meiotic Transcriptome Using Isolated Meiocytes of *Arabidopsis thaliana*

Changbin Chen and Ernest F. Retzel

Abstract

Improved transcriptome sequencing technologies (RNA-seq) have advanced our understanding of the tissue-specific transcriptome landscapes, including those of messenger RNAs, noncoding RNAs and small RNAs. However, transcriptome profiles of plant meiocytes remain challenging due to the lack of efficient methods to enrich meiocytes for the analysis of temporal and spatial gene expression patterns during meiosis. In this chapter, we describe a method to analyze the *Arabidopsis* meiotic transcriptome using isolated male meiocytes.

Keywords RNA-seq, Meiocytes, Transcriptome, Capillary collection method, *Arabidopsis*

1 Introduction

Sequence-based differential expression (RNA-seq) has largely replaced microarrays for gene expression studies (1–8). The advantages of RNA-seq include the ability to discover novel genes and exons, alternate splicing, and short insertions and deletions (indels). Even in the completely sequenced and well-studied model plant *Arabidopsis thaliana*, this technique enabled us to discover genes in the meiotic process which were previously annotated as only "predicted," "hypothetical" or, in some cases, were completely unannotated (9).

Using flowering plants to study meiosis has some inherent methodological challenges. During reproductive development, meiosis occurs in both the male organ (anther) and the female organ (ovary). Most flowering plants bear one or more anthers in each bisexual or male flower, which make the anthers easily accessible

Wojciech P. Pawlowski et al. (eds.), *Plant Meiosis: Methods and Protocols*, Methods in Molecular Biology, vol. 990,
DOI 10.1007/978-1-62703-333-6_20, © Springer Science+Business Media New York 2013

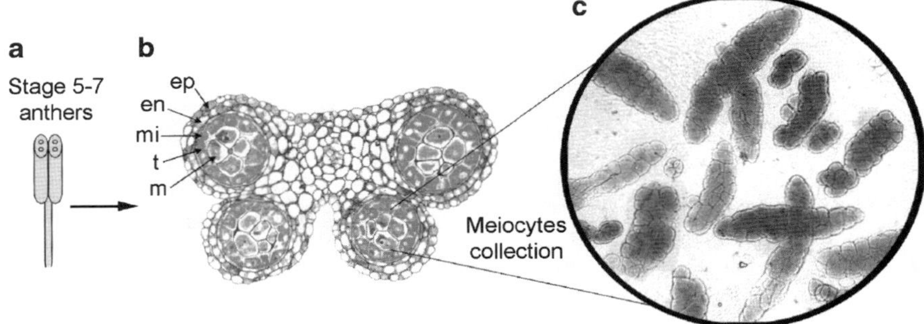

Fig. 1 *Arabidopsis* anthers and male meiocytes. (**a**) A diagram of an anther. (**b**) Micrograph of an anther cross section showing the position of meiocytes (m) and surrounding layers (*t* tapetum; *mi* middle layer; *en* endothecium; *ep* epidermis). (**c**) A view of male meiocytes collected by the Capillary Collection Method (CCM)

and thus ideal choices to study plant meiosis. The size of anthers that undergo meiosis, however, is relatively small in plants, and this is particularly true in *Arabidopsis* (10). In addition, each anther contains only a small fraction of male meiocytes (pollen mother cells, Fig. 1). For instance, male meiocytes constitute about 1% of the *Arabidopsis* anthers, making the enrichment of meiocytes very challenging but also very desirable. Thus, several methods have been developed to investigate plant meiocyte transcriptomes. They include the following: (1) Collection of anthers that are undergoing meiosis followed by transcriptomics studies, which have been performed in several species, including *Arabidopsis* (11), rice (12, 13), maize (14), and wheat (15); (2) Use of genetically ablated lines (14, 16); (3) Laser capture microdissection (LCM) of meiocytes (17); and (4) Isolation of male meiocytes from meiotic anthers (9, 18). Of all these approaches, transcriptome studies using isolated meiocytes have been the most satisfying because isolation of meiocytes has significantly eliminated the background generated by non-meiotic cells of the anthers, the bias generated by abnormal development in genetically ablated lines, and the necessity of amplification of LCM samples. In addition, the isolation of staged meiocytes is much more feasible for most laboratories, as no expensive equipment such as laser dissection microscopes is required.

In this chapter, we present the method that we use to profile meiotic transcriptomes of isolated male meiocytes. The procedure of meiocyte collection, which we named Capillary Collection Method (CCM), enables us to collect sufficient amounts of total RNA for transcriptome studies from highly concentrated meiocyte samples without the need for mRNA amplification. We used the RNA collected with CCM to conduct RNA-seq followed by bioinformatic analyses to identify meiosis-specific genes in *Arabidopsis*. We have also successfully applied this method to maize (*Zea mays*) transcriptome studies with slight modifications.

2 Materials

2.1 Microscopes

1. A stereo dissecting microscope with a 10–40× magnification.

2. An inverted microscope offering a 40–400× magnification with stage movement controls on the left side (if you are a right-handed person) or both sides, or with a motorized stage.

2.2 Capillary Collection Tubes

The capillary collection tubes are custom-made from disposable cotton-plugged borosilicate-glass Pasteur pipettes (Fisher Scientific, Suwanee, GA, USA) and brown rubber tubing (Fig. 2a). Each collection tube includes three parts: a glass Pasteur pipette used as a mouth piece, a connecting tube, and a capillary needle made from another glass pipette heated over a gas burner and by stretched into a capillary-like tube.

2.3 Collection Buffers (See Note1)

1. 1× DPBS (Dulbecco's phosphate-buffered saline) buffer: 2.7 mM KCl, 1.5 mM KH_2PO_4, 136.9 mM NaCl, 8.1 mM Na_2HPO_4, pH 7.0 (19).

2. 0.5× Murashige and Skoog (MS) medium (20): prepared using premixed MS medium powder (Sigma Aldrich, St. Louis, MO, USA) and 20% sucrose. To use for RNA extraction, add an RNAse inhibitor to a final concentration of 1 U/μl.

2.4 Meiocyte Releasing Tools

1. Dissecting needles.

2. 1 ml disposable syringes.

3. Tweezers with flat tips.

2.5 Collection Slides

1. Fisherbrand Probeon Plus slides (Fisher Scientific, Suwanee, GA, USA).

2. Regular glass slides.

3. Fisherbrand unbreakable coverslips (Fisher Scientific, Suwanee, GA, USA).

2.6 Other Reagents

1. RNAlater (Qiagen, Hilden, Germany).

2. DNA stain (e.g., DAPI or SYTOX green).

3. Ambion RNAqueous-Micro-Kit (Life Technologies, Carlsbad, CA, USA).

2.7 Equipment

1. Qubit fluorometer (Invitrogen, Carlsbad, CA, USA).

2. Agilent Bioanalyzer 2100 microfluidics (Agilent, Santa Clara, CA, USA).

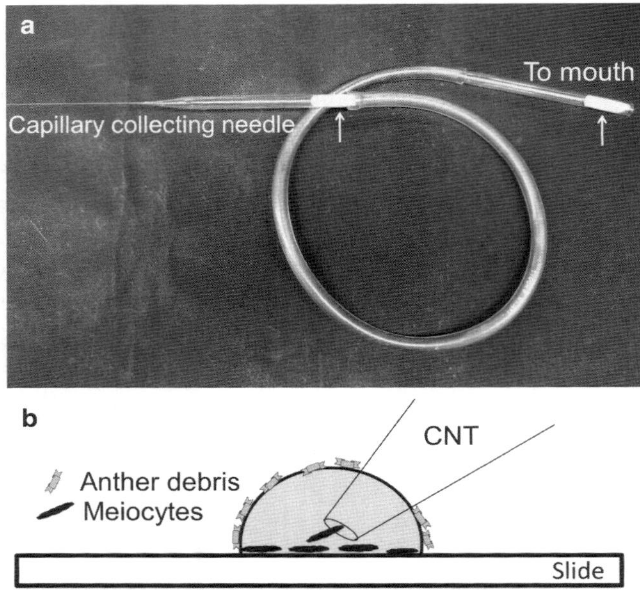

Fig. 2 The Capillary Collection Method (CCM). (**a**) A capillary collection tube. to the cotton plugs. (**b**) A close-up view of the meiocyte collection slide, showing how a charged slide surface can increase surface tension, which facilitates separation of meiocytes from other cells and cell debris. CNT = a collecting needle tip

3 Methods

3.1 Anther Collection

1. Set a dissecting microscope on a desk at a comfortable operating height.

2. Place a Fisher Probeon slide (see Note 2) on the dissecting microscope stage, add 50 µl of collection buffer onto the slide (Fig. 2b).

3. Dissect anthers from flower buds and collect 15–20 anthers into a drop of collection buffer. *Arabidopsis* anthers should be at stages 5–7 (21), which correspond to stage 9 flower buds (22). Remove any anthers that do not look transparent before proceeding to the next step (see Note 3).

3.2 Capillary Collection Method

1. Set up the inverted microscope for collecting meiocytes. The microscope should have an automatically operated stage or a stage that can be easily controlled. We have modified our microscope to be equipped with a left-hand controlled stage, which is the best setting for right-handed people.

2. Release meiocytes from anthers. For *Arabidopsis* or other plants that have small anthers, use tweezers to break the anthers. To release the meiocytes, move the tweezers—continuously squishing—through the whole buffer drop till all anthers are

broken. If anthers stick to the tweezers, use a dissecting needle to get them off back into the collection drop. Meiocytes released from anthers sink to the bottom of the buffer drop, while cell debris floats on the drop surface (Fig. 2b).

3. Transfer the slide to an inverted microscope stage. Use one hand (the right hand if you are a right-handed person) to hold the glass capillary needle and then very carefully insert the needle into the liquid drop until you touch the surface of the glass slide. Use mouth-controlled air flow through the rubber tubing to suck in meiocytes into the capillary (Fig. 2). There are two cotton plugs in the collection tube to reduce the potential of sample contamination by mouth breath (Fig. 2a). While using one hand to collect meiocytes, use the other hand to adjust the stage position to continue collecting meiocytes. In *Arabidopsis*, meiotic prophase meiocytes from each anther lobe stay together and form a cluster. Each cluster contains 20–50 meiocytes (Fig. 1c).

4. After collecting a number of cells and filling up the capillary, gently blow the cells into a microcentrifuge tube containing RNAlater. Multiple collections can be added into the same tube. The cells can be stored in RNAlater for 1 month at 4°C. Alternatively, pellet the cells for storage. To do this, add at least 1 volume of PBS to the microfuge tube (to dilute the dense RNA*later*® solution) and centrifuge at $400 \times g$ for 1 min in a microcentrifuge. The cell pellet can be then stored at –80°C.

5. To calculate the number of collected meiocytes, gently resuspend them and pipette 1 µl of the cell suspension onto a cell counting slide. Add a DNA stain, e.g., 10 µM DAPI (Vector Laboratories, Burlingame, CA, USA) or 5 µM SYTOX green (Life Technologies, Carlsbad, CA, USA), and analyze the chromosome stage with a fluorescence microscope.

3.3 RNA Extraction from Meiocyte Samples

Extract total RNA from meiocytes using the Ambion RNAqueous Micro Kit following manufacturer's instructions.

1. Add three volumes of 1× PBS to the meiocytes-RNAlater suspension (see Note 4). Centrifuge at $9,500 \times g$ for 1 min, remove the supernatant

2. Resuspend the cell pellet by vortexing vigorously in 100 µl of lysis buffer (see Note 5).

3. Wash once with Wash Solution 1 and twice with Wash Solution 2/3.

4. Elute the RNA into a Micro-Elution tube twice with 10 µl of preheated elution buffer.

5. Measure RNA yield using the Qubit fluorometer (Invitrogen, Carlsbad, CA, USA) and the Agilent Bioanalyzer 2100

microfluidics (Agilent, Santa Clara, CA, USA) instruments. On average, ~6,000 meiocytes yield 1 μg total RNA.

3.4   RNA Sequencing

Two methods have been developed for transcriptome analysis (Fig. 3). The first method captures poly-A-containing RNA, including mRNA and long mRNA-like noncoding RNA (23). The second technique, referred to as "whole transcriptome RNA sequencing," captures mRNA as well as noncoding RNA longer than 200 bp. Expression data obtained with the second method are sometimes confounded by intron sequences, as unprocessed non-polyadenylated pre-mRNAs are captured as well.

3.4.1   Poly-A RNA Sequencing

1. Assess integrity of the extracted RNA using Bioanalyzer 2100 following manufacturer's protocol.

2. Purify the poly-A containing RNA fraction from 2.5 μg of total RNA with poly(T) coupled magnetic beads using two rounds of purification.

3. Fragment the eluted poly-A RNA for more efficient first-strand synthesis priming.

4. Construct the cDNA library using the TruSeq RNA Kit (Illumina, Inc., San Diego, CA, USA) following manufacturer's protocol.

5. Check the quantity and quality of the library DNA with the NanoDrop and Bioanalyzer instruments. Verify the DNA quantification using qPCR (see Note 6).

3.4.2   Whole Transcriptome RNA Sequencing

1. Assess integrity of the extracted RNA using Bioanalyzer 2100 following manufacturer's protocol.

2. Fragment 200 ng of total RNA at an elevated temperature.

3. Construct the cDNA library using the TruSeq RNA Kit (Illumina, Inc., San Diego, CA, USA) following manufacturer's protocol.

4. Check the quantity and quality of the library DNA using the NanoDrop and Bioanalyzer instruments.

5. Perform Duplex-Specific thermostable Nuclease (DSN) normalization of the library using 100 ng of the library DNA. To do this, follow the appropriate Illumina protocol.

6. Check the quantity and quality of the library DNA with the NanoDrop and Bioanalyzer instruments. Verify DNA quantification using qPCR (see Note 6).

3.5   Transcriptome Data Analysis

3.5.1   Aligning the Reads to the Arabidopsis Reference Genome

1. Align sequencing reads to the *Arabidopsis* genome (TAIR 10). Several methods can be used for this purpose. We use the GSNAP algorithm (24) and the Alpheus software system (25). GSNAP is parameterized to allow a single indel within a read alignment in order to determine spliced read alignments inde-

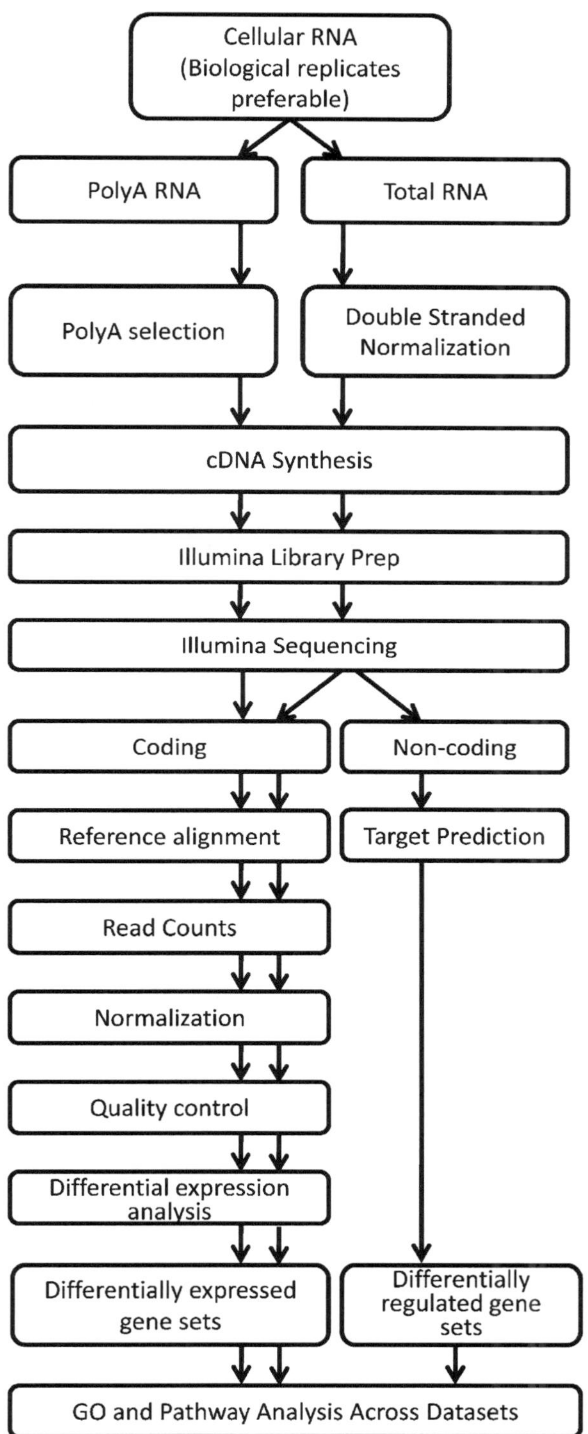

Fig. 3 Flowchart of meiotic transcriptome analysis

pendently of the exon junctions given in the *Arabidopsis* genome annotation. Intron lengths are restricted to no larger than 10 kb.

2. Uniquely aligned reads are counted as representing TAIR 10 gene models as long as the read alignment overlaps the gene. However, we allow the possibility of novel exons and small UTR extensions.

3. Reads aligning to multiple genome locations (more than one but fewer than 5 locations) are fractionally apportioned to all these locations using an iterative procedure similar to the rescue method described by Mortazavi et al. (26). However, we use the count values established for less ambiguous reads to incrementally update read counts before using the updated counts to determine the apportioning ratios for the next higher level of mapping ambiguity. The apportioned reads are then normalized for differences in sequencing depth using the trimmed mean of M-values (TMM) method (27) implemented in EdgeR (28). rRNA genes are excluded from this procedure to further control for the impact of differences in efficacy of rRNA reduction methods on the effective library sizes.

4. For comparing of expression between genes within a single library, we normalize counts for transcript length differences using the maximum length among all splice isoforms of a given gene. Candidate noncoding alignments are evaluated using ESTScan2 (29).

3.5.2 Expression Analysis

1. We perform all gene expression and related analyses with the JMP Genomics software (SAS Institute, Cary, NC, USA) (25). Normalized read counts are log10+1 transformed. Quality control and exploration of data are performed through overlaid kernel density estimates, univariate distribution results, Mahalanobis distances, correlation coefficients of pairwise sample comparisons, unsupervised principal component analysis (by Pearson product–moment correlation), and Ward hierarchical clustering of Pearson product–moment correlations of read frequencies.

2. In our expression analyses, we use decomposition of principal components of variance of normalized read counts and false discovery rate (FDR)-corrected analysis of variance to determine the drivers of variance and identify differentially regulated gene sets.

3. We perform gene ontology (GO) analysis using Blast2GO, and pathway analysis—using GeneGO.

4. *P*-values for candidate differentially expressed genes are determined for each pairwise comparison between samples using Fisher's exact test against the contingency table formed from

the apportioned read counts for each gene and effective library sizes represented by the total aligned read counts (but excluding reads aligning to rRNA genes); false discovery rate (FDR) adjustments were made to correct for multiple testing using the Benjamini–Hochberg procedure (30).

5. The lack of biological replication within the experimental design limits the degree to which statistically significant differences between samples can be interpreted as biologically significant, since the level of biological variability for the given gene may be high. As an approximation to estimate biological variability in the absence of true biological replicates, we use the DESeq package (31) to fit an estimate of variance from the data as a function of expression magnitude and to perform differential expression testing using a conditioned test procedure analogous to Fisher's test.

4 Notes

1. Either of these two collecting buffers can be used for *Arabidopsis* meiocytes collection.

2. We use the surface tension phenomenon to separate meiocytes from anther wall cells/debris. Meiocytes sink to the bottom of the buffer drop, while tissue floats on the surface. Therefore, use slides that generate strong surface tension rather than regular microscope slides without a charged surface. Charged slides such as Fisherbrand Probeon Plus (Fisher Scientific, Suwanee, GA, USA) can be replaced with regular glass slides covered with Fisherbrand plastic unbreakable cover slips, which are less expensive but also create surface tension.

3. Anthers undergoing meiosis are transparent. If anther lobes are not transparent, this indicates that the meiocytes have already completed meiosis. For example, when meiocytes are at the tetrad stage, anther lobes appear off-white. Yellowish anther lobes contain pollen grains. Microspores and pollen grains cannot be easily separated from meiocytes after they are released from anthers as they also sink to the bottom of the collection buffer drop.

4. Meiocytes in RNA*later*® are very hard to pellet, even at the maximum microcentrifuge speed. To facilitate meiocyte sedimentation, add 1× PBS to the meiocyte suspension before centrifugation.

5. In order to recover all RNA types, including mRNAs, noncoding RNAs, and small RNAs, add 125 µl of 100% ethanol.

6. Accurate quantification of library DNA is needed for creating optimum cluster densities, which is essential for achieving high quality of Illumina sequencing data.

Acknowledgments

This work was supported by the Biotechnology Research and Development Corporation (BRDC). The application of this method to other plant species, such as maize, is supported by the National Science Foundation (IOS-1025881). Technical support has been provided by Stefanie Dukowic-Schulze, Marie-Therese Kurzbauer, Andrew Farmer, Ingrid Lindquist, and Joann Mudge.

References

1. Malone JH, Oliver B (2011) Microarrays, deep sequencing and the true measure of the transcriptome. BMC Biol 9:34

2. Fox-Walsh K, Davis-Turak J, Zhou Y, Li H, Fu XD (2011) A multiplex RNA-seq strategy to profile poly(A(+)) RNA: Application to analysis of transcription response and 3′ end formation. Genomics 98:266–271

3. Graveley BR, Brooks AN, Carlson JW, Duff MO, Landolin JM, Yang L et al (2011) The developmental transcriptome of Drosophila melanogaster. Nature 471:473–479

4. Ozsolak F, Milos PM (2011) RNA sequencing: advances, challenges and opportunities. Nat Rev Genet 12:87–98

5. Oshlack A, Robinson MD, Young MD (2010) From RNA-seq reads to differential expression results. Genome Biol 11:220

6. Majewski J, Pastinen T (2011) The study of eQTL variations by RNA-seq: from SNPs to phenotypes. Trends Genet 27:72–79

7. Costa V, Angelini C, De Feis I, Ciccodicola A (2010) Uncovering the complexity of transcriptomes with RNA-Seq. J Biomed Biotechnol 2010:853916

8. Nacu S, Yuan W, Kan Z, Bhatt D, Rivers CS, Stinson J et al (2011) Deep RNA sequencing analysis of readthrough gene fusions in human prostate adenocarcinoma and reference samples. BMC Med Genomics 4:11

9. Chen C, Farmer AD, Langley RJ, Mudge J, Crow JA, May GD et al (2010) Meiosis-specific gene discovery in plants: RNA-Seq applied to isolated Arabidopsis male meiocytes. BMC Plant Biol 10:280

10. Sanchez-Moran E, Mercier R, Higgins JD, Armstrong SJ, Jones GH, Franklin FCH (2005) A strategy to investigate the plant meiotic proteome. Cytogenet Genome Res 109:181–189

11. Wijeratne AJ, Zhang W, Sun Y, Liu W, Albert R, Zheng Z et al (2007) Differential gene expression in Arabidopsis wild-type and mutant anthers: insights into anther cell differentiation and regulatory networks. Plant J 52:14–29

12. Huang MD, Wei FJ, Wu CC, Hsing YI, Huang AH (2009) Analyses of advanced rice anther transcriptomes reveal global tapetum secretory functions and potential proteins for lipid exine formation. Plant Physiol 149:694–707

13. Wang Z, Liang Y, Li C, Xu Y, Lan L, Zhao D et al (2005) Microarray analysis of gene expression involved in anther development in rice (*Oryza sativa* L.). Plant Mol Biol 58:721–737

14. Ma J, Duncan D, Morrow DJ, Fernandes J, Walbot V (2007) Transcriptome profiling of maize anthers using genetic ablation to analyze pre-meiotic and tapetal cell types. Plant J 50:637–648

15. Crismani W, Baumann U, Sutton T, Shirley N, Webster T, Spangenberg G et al (2006) Microarray expression analysis of meiosis and microsporogenesis in hexaploid bread wheat. BMC Genomics 7:267

16. Wijeratne AJ, Ma H (2007) Genetic analyses of meiotic recombination in *Arabidopsis.* J Integ Plant Biol 49:1199–1207

17. Tang X, Zhang ZY, Zhang WJ, Zhao XM, Li X, Zhang D et al (2010) Global gene profiling of laser-captured pollen mother cells indicates molecular pathways and gene subfamilies involved in rice meiosis. Plant Physiol 154:1855–1870

18. Yang H, Lu P, Wang Y, Ma H (2011) The transcriptome landscape of Arabidopsis male meiocytes from high-throughput sequencing: the complexity and evolution of the meiotic process. Plant J 65:503–516

19. Dulbecco R, Vogt M (1954) Plaque formation and isolation of pure lines with poliomyelitis viruses. J Exp Med 99:167–182

Murashige T, Skoog F (1962) A revised medium for rapid growth an dbio-assay wth tobacco tissue cells. Physiol Plant 15:473–497

21. Sanders PM, Bui AQ, Weterings K, McIntyre KN, Hsu YC, Lee PY et al (1999) Anther development defects in Arabidopsis thaliana male-sterile mutants. Sex Plant Reprod 11:25

22. Smyth DR, Bowman JL, Meyerowitz EM (1990) Early flower development in Arabidopsis. Plant Cell 2:755–767

23. Rymarquis LA, Kastenmayer JP, Huttenhofer AG, Green PJ (2008) Diamonds in the rough: mRNA-like non-coding RNAs. Trends Plant Sci 13:329–334

24. Wu TD, Nacu S (2010) Fast and SNP-tolerant detection of complex variants and splicing in short reads. Bioinformatics 26:873–881

25. Miller NA, Kingsmore SF, Farmer A, Langley RJ, Mudge J, Crow JA et al (2008) Management of high-throughput DNA sequencing projects: Alpheus. J Comput Sci Syst Biol 1:132

26. Mortazavi A, Williams BA, McCue K, Schaeffer L, Wold B (2008) Mapping and quantifying mammalian transcriptomes by RNA-Seq. Nat Methods 5:621–628

27. Robinson MD, Oshlack A (2010) A scaling normalization method for differential expression analysis of RNA-seq data. Genome Biol 11:R25

28. Robinson MD, McCarthy DJ, Smyth GK (2010) edgeR: a Bioconductor package for differential expression analysis of digital gene expression data. Bioinformatics 26:139–140

29. Iseli C, Jongeneel CV, and Bucher P (1999) ESTScan: a program for detecting, evaluating, and reconstructing potential coding regions in EST sequences. Proc Int Conf Intell Syst Mol Biol 138–148.

30. Benjamini Y, and Hochberg Y (1995) Controlling the false discovery rate: a practical and powerful approach to multiple testing. J Royal Stat Soc. Series B (Methodological) 57(1): 289–300.

31. Anders S, Huber W (2010) Differential expression analysis for sequence count data. Genome Biol 11:R106

Chapter 21

Analysis of Meiotic Protein Complexes from *Arabidopsis* and *Brassica* Using Affinity-Based Proteomics

Kim Osman, Elisabeth Roitinger, Jianhua Yang, Susan Armstrong, Karl Mechtler, and F. Chris H. Franklin

Abstract

The application of proteomics techniques to the study of plant meiosis has the potential to make a valuable contribution to our understanding of the molecular events underpinning meiotic processes. Here we describe the preparation of meiotic protein complexes from *Arabidopsis thaliana* and its close crop relative, *Brassica oleracea*, by co-immunoprecipitation for in-solution analysis by tandem mass spectrometry (MS/MS). Early results using these techniques have proved encouraging, enabling the identification of candidate AtASY1-interacting proteins in *A. thaliana* and providing evidence of an *in planta* interaction between BoASY1 and BoASY3 in *B. oleracea*. The detection of phospho-modified peptides of BoASY1 and BoASY3 suggests that this approach may be useful for studying meiotic protein modification events.

Keywords Meiosis, Co-immunoprecipitation, Mass spectrometry, Protein complexes, *Arabidopsis*, *Brassica*

1 Introduction

Meiosis provides the basis for the generation of new traits of agronomic, environmental and economic importance through plant breeding. Recent years have seen substantial progress in our understanding of some of the molecular mechanisms involved in this process (reviewed in ref. 1), yet many challenges remain. In the past, successful identification of plant meiotic genes has often involved the use of mutant screens or *in silico* orthologue searches. However, these approaches have limitations: meiotic mutations may have only a subtle effect on fertility and thus be overlooked in mutant screens; many plant genes belong to multi-gene families

---

*Kim Osman and Elisabeth Roitinger have contributed equally to this work.

Wojciech P. Pawlowski et al. (eds.), *Plant Meiosis: Methods and Protocols*, Methods in Molecular Biology, vol. 990, DOI 10.1007/978-1-62703-333-6_21, © Springer Science+Business Media New York 2013

215

and exhibit redundancy; and functional orthologues, particularly those encoding components of the SC, may be poorly conserved at the primary sequence level. To address these issues, an alternative method of meiotic gene identification based on the use of proteomics was developed (2). In an extension of this approach, we are using antibodies raised against known *Arabidopsis thaliana* meiotic proteins to immunoprecipitate complexes from *A. thaliana* and its close crop relative, *Brassica oleracea*, for in-solution analysis by tandem mass spectrometry (MS/MS). In addition to identifying more meiotic proteins, we anticipate that co-immunoprecipitation will be a powerful tool for investigating binary and higher order protein–protein interactions. We believe that this approach is timely given the accumulating evidence from other model organisms that key meiotic events are regulated at the protein modification level (3, 4).

Analyzing complexes from both *Arabidopsis* and *Brassica* allows us to harness the particular advantages of each species in a complementary manner. *Arabidopsis* has comprehensive, well-annotated databases and extensive mutant collections but, due to its small bud size, is a relatively poor source of meiotic tissue and large scale meiotic staging (for meiotic enrichment of samples) is only practical based on bud size. *B. oleracea* buds, on the other hand, are much bigger making it feasible to stage individual buds at the cytological level. Furthermore, routine harvesting of anthers or even meiocytes is practicable, leading to considerably improved meiotic enrichment in tissue samples. However, *Brassica* does suffer from a number of disadvantages compared to *Arabidopsis*: it has a complex genome (resulting from a triplication event which occurred subsequent to its evolutionary divergence from *Arabidopsis*), there are currently few publicly available *Brassica* meiotic mutants, and sequence databases are limited, although this situation will improve following the publication of the *Brassica rapa* genome sequence (5).

In this chapter we describe the selection and preparation of meiotic tissue from *A. thaliana* and *B. oleracea*, the extraction and co-immunoprecipitation of meiotic protein complexes and the analysis of complexes by tandem mass spectrometry (MS/MS). We have recently used these methods to demonstrate an *in planta* interaction between the axis-associated protein, BoASY1 and BoASY3 (6) and to identify potential ASY1-interacting proteins in *A. thaliana*. Significantly, we have also been able to identify a number of phospho-modified peptides in BoASY1 and BoASY3, demonstrating the potential of this approach for future protein modification studies.

2 Materials
(See Note 1)

2.1 Plant Growth and Harvesting

1. *Arabidopsis thaliana* accession Columbia-0 and *B. oleracea* var. *alboglabra* A12DHd grown in multipurpose compost in a glasshouse at 21°C with supplementary lighting (16 h light, 8 h dark).

2. Phosphate Buffered Saline (PBS): 0.16 M NaCl, 3 mM KCl, 8 mM Na_2HPO_4, 1 mM KH_2PO_4, pH 7.3.

3. Protease inhibitor cocktail: dissolve one Complete mini EDTA-free tablet (Roche Applied Science, Penzberg, Germany) in 10 ml of PBS. Store at −20°C in 500 µl aliquots. Once thawed, any inhibitor remaining after use should be discarded.

4. Aceto-orcein stain: dissolve 0.5 g of orcein in 5 ml lactic acid plus 5 ml propionic acid. Dilute 1:1 with water before use.

5. Watchmakers' forceps (two pairs).

6. Cavity microscope slide.

7. Liquid nitrogen.

2.2 Protein Extraction and Co-immunoprecipitation

1. IP-buffer: 20 mM Tris–HCl (pH 7.5), 150 mM NaCl, 10% glycerol, 2 mM EDTA.

2. NP-40: Nonidet P40.

3. Protease inhibitor cocktail: Complete mini EDTA-free tablets (Roche Applied Science, Penzberg, Germany). Use 1 tablet per 10 ml IP-buffer. Prepare immediately prior to use.

4. Phosphatase inhibitor cocktail: Halt Phosphatase Inhibitor Cocktail (Thermo Fisher Scientific, Waltham, MA, USA). Add 100 µl per 10 ml IP-buffer immediately before use.

5. TBST: 20 mM Tris–HCl (pH 7.5), 150 mM NaCl, 0.04% Triton X-100.

6. Na-borate: 0.2 M, pH 9.2 (see Note 2).

7. DMP solution: 5.2 mg/ml DMP (dimethyl pimelimidate dihydrochloride; Sigma Aldrich, St. Louis, MO, USA) in 0.2 M Na-borate, pH 9.2 (see Note 3).

8. Tris–HCl: 0.2 M, pH 8.0.

9. Tris–HCl: 1.5 M, pH 9.2.

10. Glycine: 0.1 M, pH 2.0.

11. NaN_3 solution: Prepare a 20% stock (200 mg/ml) in water. Store at 4°C.

12. NaCl: 150 mM.

13. Protein A beads: Affi-Prep Protein A (Bio-Rad, Hercules, CA, USA) (see Note 4).

14. Antibodies: antibody specific to the protein of interest and non-specific IgG or antibodies unrelated to the protein of interest (see Note 5).

15. Low retention 0.2 ml PCR tubes (Axygen, Union City, CA, USA).

2.3 Protein Digestion and MS-Measurement of the Generated Peptides

1. ABC solution: 100 mM ammonium bicarbonate.

2. DTT solution: 1 mg/ml DTT (1,4-dithiothreitol) in 100 mM ABC.

3. MMTS solution: prepare a 200 mM stock of S methyl methanethiosulfonate (Sigma Aldrich, St. Louis, MO, USA) in isopropanol. Store at –20°C. Dilute to 28 mM in 100 mM ABC (1:7) directly before use.

4. Trypsin: prepare a 100 ng/µl stock of Trypsin Gold, mass spectrometry grade (Promega, Madison, WI, USA) in 50 mM acetic acid. Store at –80°C.

5. TFA: trifluoroacetic acid (Thermo Fisher Scientific, Waltham, MA, USA).

6. FA: formic acid, Suprapur 98–100% (Merck, Whitehouse Station, NJ, USA).

7. Glacial acetic acid.

8. TFE: trifluoroethanol, ReagentPlus 99% (Sigma Aldrich, St. Louis, MO, USA).

9. ACN: acetonitrile, LC-MS, Chromasolv (Sigma Aldrich, St. Louis, MO, USA).

10. Solvent A: 0.1% FA, 5% ACN.

11. Solvent B: 0.08% FA, 30% ACN.

12. Solvent C: 0.08% FA, 80% ACN, 10% TFE.

3   Methods

To reduce the risk of false identification of potential interacting proteins, particularly proteins represented by relatively few peptides, we suggest carrying out at least three replicates of co-immunoprecipitation (Co-IP) experiments.

3.1 Staging and Harvesting Brassica Anthers and Meiocytes

1. Choose inflorescences prior to flower opening. Working with one inflorescence at a time, use a dissecting microscope with halogen lighting to dissect out individual buds and place them in size-order in a Petri dish containing damp filter paper.

2. To stage a bud, carefully open sepals and remove one of the six anthers by excising it at the base with forceps. Transfer the anther to a drop of aceto-orcein stain on a glass slide and

add a cover slip, pressing down gently. Use phase-contrast microscopy to determine meiotic stage (see Note 6).

3. Transfer buds of the selected stage (we typically use buds in prophase I) to damp filter paper on ice while staging remaining buds in inflorescence (see Note 7).

4. Dissect remaining anthers from selected buds, transfer to a microfuge tube and quickly freeze in liquid nitrogen. Store at −80°C (see Note 8). If meiocytes are required, do not freeze anthers but continue with steps 5–8.

5. Place 20 µl of chilled PBS containing the protease inhibitor cocktail into a cavity slide under a dissecting microscope.

6. Transfer staged anthers to the edge of the well. Using forceps, pierce each anther tip in turn and apply gentle pressure to extrude meiocytes into the liquid in the well. If required, "empty" anthers can be retained as a negative control. Do not allow meiocyte suspension to dry out; add more PBS solution in 10 µl aliquots as required. Transfer no more than 20–30 anthers to the slide at a time and keep the remainder on ice until required.

7. At intervals, use a pipette tip to transfer meiocyte suspension to a microfuge tube and freeze in liquid nitrogen. Meiocytes from several sessions can be pooled in the same tube (see Note 8).

3.2 Harvesting Arabidopsis Meiotic Buds

Although it is not practical to routinely stage individual *Arabidopsis* buds, reasonable care is taken to select meiotic buds by dissecting inflorescences using a microscope and choosing buds of an appropriate size. Until experienced, anthers or whole buds should be stained and examined (*as in* Subheading 3.1) in order to become familiar with the required bud size.

1. For collecting meiotic buds, choose inflorescences that are just beginning to open or earlier.

2. Dissect inflorescences and collect buds on moist filter paper.

3. At intervals of no more than 30 min freeze buds in liquid nitrogen, pooling them (see Note 8).

4. After harvesting, prune plants to produce more flower stalks. Following three rounds of harvesting, plants should be discarded due to reduced quality of bud material.

3.3 Cross-linking Antibodies to Beads

Unless otherwise stated, perform all steps on ice or at 4°C.

1. We find it convenient to cross-link beads in 100 µl aliquots. Resuspend beads in storage buffer and quickly transfer required volume to a 1.5 ml microfuge tube (it will be necessary to resuspend bead stock between each transfer).

2. Equilibrate beads by washing 2× with 10 bed-volumes (bv) of TBST. Resuspend beads by flicking tubes, pellet them by spinning

for 10 s in a microfuge tube and remove supernatant (SN) using a fine pipette tip or syringe with a 27 G needle.

3. Add 10 bv TBST to the washed beads and add purified antibody to the required concentration (we recommend the use of 1 mg antibody per ml beads initially). Mix beads with rotation for 1 h at room temperature or 4°C overnight.

4. Wash beads 3× briefly with 10 bv of TBST then 3× briefly with 10 bv of Na-borate solution at room temperature.

5. Add 20 bv of DMP solution to the beads and mix with rotation at room temperature for 30 min (do not mix for longer).

6. Stop the cross-linking reaction by washing beads 2× with 10 bv of 0.2 M Tris–HCl pH 8.0 for 10 min.

7. Wash beads briefly with 10 bv of TBST to re-equilibrate.

8. To remove non-cross-linked antibody, wash beads 2× briefly with 10 bv of glycine solution then 3× briefly with 10 bv of TBST to re-equilibrate. Store antibody-coupled beads in TBST plus 0.05% NaN_3 at 4°C until required.

9. Before use, remove storage buffer and wash 2× briefly with TBST.

3.4 Protein Extraction and Co-immunoprecipitation

For a typical Co-IP experiment we use approximately 1 ml of *Arabidopsis* meiotic buds or staged *Brassica* anthers (or meiocytes) from 200 buds. As the stability of meiotic protein complexes in storage is largely unknown, extracts are used immediately for Co-IP and not frozen. The Co-IP procedure is based on a method described by Herzog and Peters (7). Unless otherwise stated, perform all steps on ice or at 4°C.

1. Powder plant tissue by grinding in liquid nitrogen for 10 min using a small prechilled mortar and pestle.

2. Resuspend the tissue in 1 ml IP-buffer with 0.1% NP-40 and protease/phosphatase inhibitors added. Transfer slurry to 2 ml microfuge tubes on ice and use an additional 1 ml buffer to collect any residue from mortar and pestle.

3. Clarify extracts by spinning at 14,000×*g* for 10 min. Transfer SN to fresh tubes and repeat spin. Pool SN into a single tube and remove a 100 µl aliquot which can be used for assaying protein yield, analysis by SDS-PAGE, immunoblotting etc. if required.

4. Preclear remaining extract using 100 µl of beads coupled to non-specific IgG. Incubate with rotation for 30 min, pellet the beads and divide the SN equally between two fresh tubes.

5. Add 50 µl of specific antibody-coupled beads to one of the tubes; to the other add 50 µl of beads coupled to the control antibody (non-specific for the protein of interest) (see Note 5). Incubate with rotation for 30 min, pellet the beads and remove the SN (retain for analysis if required).

6. Wash beads 3× with 2 ml (40 bv) of ice-cold IP-buffer containing 0.1% NP-40 and protease/phosphatase inhibitors. Resuspend beads, incubate with rotation for 5 min, pellet beads and remove as much SN as possible using a needle (see Note 9).

7. Wash beads 4× with 2 ml of IP-buffer alone (see Note 10).

8. Wash beads briefly with 2 ml of 150 mM NaCl to remove pH-buffering agents. Resuspend beads in 150 mM NaCl to allow transfer to 0.5 ml microfuge tubes. Pellet beads and remove as much SN as possible.

9. Elute proteins by adding 50 μl (1 bv) of glycine solution to the beads. Mix by flicking tube, pellet beads and transfer SN to a fresh Axygen low-retention 0.2 ml PCR tube containing 2.5 μl of 1.5 M Tris–HCl, pH 9.2. Repeat elution once (see Notes 11 and 17).

10. Test the pH of neutralized eluates by spotting 1 μl onto pH paper. They should be at approximately pH 8.0. If they are too acidic, add 1 μl of 1.5 M Tris–HCl pH 9.2, mix and retest. Continue until a pH of ~8.0 is reached.

11. Transfer 5 μl (1/10th volume) of eluates to fresh 0.5 ml microfuge tubes for gel analysis if required (see Note 12). Freeze remaining eluate in liquid nitrogen and store at −80°C or use directly for MS/MS sample preparation.

3.5 Protein Digestion and Measurement by HPLC-MS/MS

The protein digest method and HPLC-MS/MS setup we describe was successfully applied to identify potential ASY1-interacting proteins in *A. thaliana* and to demonstrate an *in planta* interaction between ASY1 and ASY3 and for phospho-analysis in *B. oleracea*. Other labs may wish to use alternative setups (see Note 13).

1. Depending on the results of the protein gel analysis (see Subheading 3.4, step 11 and Note 12), the two consecutive eluates may be pooled. Reduce the proteins in the pooled eluate (90 μl) by incubation with 4 μl of a fresh DTT-solution at 56°C for 30 min.

2. Alkylate the reduced proteins by incubation with 4 μl of MMTS (28 mM solution) at room temperature for 30 min.

3. Digest the proteins in solution by incubation with 200 ng of trypsin for 4 h at 37°C and then with an additional 200 ng of trypsin overnight.

4. Acidify the protein digests by addition of 20 μl of 10% TFA.

5. Separate the generated peptides by nano-reversed phase HPLC (see Note 14).

6. For sample cleanup and pre-concentration, load the sample (up to 100 μl) on a reversed phase trap column (C18, 300 μm ID × 5 mm length, particle size 5 μm, pore size 100 Å) using 0.1% TFA at a flow rate of 25 μl/min.

7. After washing the bound peptides on the trap column, it is switched in series with an analytical reversed phase separation column (C18, 75 µm ID × 250 mm length, particle size 3 µm, pore size 100 Å), which is equilibrated in Solvent A. Bound peptides are eluted at a flow rate of 275 nl/min by applying a linear gradient from 0 to 100% Solvent B. Adjust the steepness of the gradient to the complexity of the sample. The first gradient is followed by a second steep gradient to 85% Solvent B and 15% Solvent C and a third steep gradient to 10% Solvent B and 90% Solvent C. After a wash step at 90% Solvent C, the analytical column is re-equilibrated with Solvent A and the trap column is switched off-line again (see Note 15).

8. The peptides eluting from the reversed phase column are analyzed online by nano-ESI-MS/MS. A sensitive and robust MS setup which is currently installed in many MS-laboratories performs peptide analysis on an LTQ-Orbitrap mass spectrometer (Thermo Fisher Scientific, Waltham, MA, USA) operating in data-dependent mode: each full scan acquired in the Orbitrap with high resolution and accurate mass is followed by up to 20 MS/MS scans of the most intensive ions in the ion trap with high scan speed but at low resolution. If phosphorylated peptides are expected to be present, MS/MS spectra are acquired with multistage activation enabled for neutral loss of phosphoric acid (32.66, 48.99, and 97.97). Use dynamic exclusion for data acquisition with exclusion duration for 90 s and an exclusion mass width of 5 ppm.

9. The control sample and the antibody-specific sample have to be analyzed in consecutive HPLC-MS/MS runs applying the same parameters. Start with the analysis of the control sample then perform a wash run to control for possible carryover before analyzing the antibody-specific sample.

10. Interpret the recorded MS/MS-spectra using an MS/MS database search engine, such as Mascot (Matrix Science, London, UK) or Sequest (Thermo Fisher Scientific, Waltham, MA, USA), both of which can be used in conjunction with the Proteome Discoverer Software package (Thermo Fisher Scientific, Waltham, MA, USA).

11. To identify peptides and proteins from *A. thaliana* samples perform a search against The Arabidopsis Information Resource database (TAIR10_pep_20101214; ftp://ftp.arabidopsis. org/home/tair/Proteins/TAIR10_protein_lists/) supplemented with common proteomics contaminants. For *B. oleracea*, search against the *Brassica rapa* Genome Sequencing Project Consortium database (Brapa_gene_v1.1.pep.tar.gz; http://brassicadb.org/brad/downloadOverview.php?PHPSE SSID=2um3t7bi38c56dbbhj9pmbc3b1) (5) (see Note 16).

12. Use the following search parameters: allow a precursor mass tolerance of 5 ppm, a fragment ion mass tolerance of 0.5 Da, allow two missed cleavage sites for trypsin, specify modification of cysteine with methyl methanethiosulfonate as a fixed modification, oxidation of methionine, and, in the case of a search for phosphorylated residues, phosphorylation on serine, threonine, and tyrosine as a variable modification.

13. To compare the identified peptides from the control versus the specific purification and to distinguish nonspecifically binding proteins from real interactors, the Scaffold software version Scaffold 3 (Proteome Software, Portland, OR, USA) can be used. Load the Mascot.dat files into Scaffold and set the following minimum acceptance thresholds for the search results: define the Mascot Ions Score cutoff as 20 and define a minimum of two identified unique peptides per protein to positively identify the protein. Proteins that are identified in three replicates of the specific sample, but not in three replicates of the control sample, can be considered as potential interactors (see Note 17).

4   Notes

1. Due to the high sensitivity of MS analysis, during sample preparation it is very important to minimize contamination from extraneous proteins, particularly keratin (from dead skin cells), which is a major component of lab dust. Solutions should be made up using fresh, analytical grade reagents, MilliQ or other high quality water (we use BPC grade from Sigma Aldrich, St. Louis, MO, USA), and clean equipment. Nitrile gloves should be worn throughout.

2. To dissolve Na-borate crystals, heat in a microwave and cool to room temperature before use.

3. DMP is moisture sensitive. Remove stock from −20°C freezer 30 min before use to allow equilibration to room temperature before opening. DMP solution should be freshly prepared before each experiment.

4. All antibodies used in our experiments were raised in rabbits. For antibodies raised in other organisms, it may be more appropriate to use Protein G beads (for example, GammaBind Plus Sepharose, GE Healthcare, Little Chalfont, UK). Consult relevant manufacturers' literature for relative binding capacities.

5. Success of the technique will depend on antibody quality and specificity. We recommend using affinity-purified antibodies whenever possible. Non-specific IgG or antibodies unrelated to the protein of interest are required for preclearing protein samples and to provide a negative Co-IP control.

6. As a rough guide, *B. oleracea* meiotic buds are approximately 2 mm in length and at the onset of meiosis anther tips develop a reddish pigmentation. A useful feature of both *B. oleracea* and *A. thaliana* is their high degree of meiotic synchrony such that the meiocytes within all the anthers of an individual flower bud tend to be at the same stage of meiosis.

7. To maintain integrity of subsequent protein extracts, minimize disruption to the bud and the remaining five anthers and keep them cool to reduce risk of degradation by proteases.

8. It will usually be necessary to collect tissue over several sessions. To avoid a situation where a small amount of material is spread over a large number of tubes, later collections can be added to the same tube providing the original tissue is not allowed to thaw. We suggest transferring the collection tube from the freezer directly into liquid nitrogen and only removing it for a few seconds to add the new material to the side of the upper part of the tube.

9. If necessary, in addition to the three standard washes, higher-stringency washes using higher salt and detergent concentrations (for example, 0.45 M NaCl, 1% Triton X-100) can be performed. However, to date we have only carried out *Arabidopsis/Brassica* Co-IPs using standard washes.

10. It is very important to remove all traces of detergent from samples as it interferes with MS/MS analysis. Protease inhibitors are also omitted from final washes to avoid inhibition of trypsin during MS/MS sample preparation.

11. If quantification (using iTRAQ) of the immunoprecipitated proteins is required, the proteins have to be eluted with 50 mM HCl instead of glycine and the eluates have to be neutralized with 1 M triethyl ammonium bicarbonate buffer instead of Tris.

12. We usually use SDS-PAGE followed by silver-staining to assess the amount of protein in eluates before MS/MS analysis.

13. Different strategies for protein digestion and peptide measurement by HPLC-MS/MS can be applied to successfully analyze affinity-purified complexes.

    Purified proteins can be separated by SDS-PAGE, from which either lanes covering the whole separation range, or stained protein bands specifically detected in the samples of interest, are cut out and subjected to in-gel protein digestion. Alternatively, the purified proteins can be directly digested in solution, if they are contained in a buffer compatible with the enzymatic digest. Depending on the chemical properties of the purified proteins, an appropriate enzyme is chosen for the proteolytic digest, with trypsin being in most cases the preferred enzyme. The generated peptide sample has to be desalted and concentrated either off-line via reversed phase tips or online on a reversed phase trap column, which is integrated into the nano-HPLC setup. After the desalting step, peptides are separated on an analytical reversed phase nano-HPLC column which, in most cases, will be coupled online to a mass spectrometer via a nano-electrospray ion source. Alternatively, HPLC fractions can be collected for spotting onto a MALDI-target for subsequent analysis by MALDI-MS/MS.

14. For the *B. oleracea* and *A. thaliana* Co-IP experiments we have used an Ultimate 3000 nano-LC system (Dionex, Sunnyvale, CA, USA) with PepMap C18 trap and analytical columns. The HPLC system was directly coupled to an LTQ Orbitrap Velos mass spectrometer (Thermo Fisher Scientific, Waltham, MA, USA) via a nano-electrospray ion source (Proxeon, Copenhagen, Denmark). Any other established nano-HPLC-MS/MS system can be used for the separation and analysis of proteolytic peptides.

15. If the separation system allows the use of only two solvents, prepare solvent B with 80% ACN with or without addition of 10% TFE and adjust the gradient to reach 30% ACN.

16. The *Brassica rapa* Genome Sequencing Project Consortium data is currently the best available database for carrying out searches of *B. oleracea* peptides when there is no prior expectation of protein presence. When analyzing known proteins of interest, however, higher sequence coverage "scores" may be achieved by adding the relevant protein sequences into the database prior to searching.

17. As an alternative, quantitative proteomics methods can be employed by including a differential isotope labeling step such as dimethyl labeling, iTRAQ, or TMT, which enable the measurement of the relative abundance of all proteins present in the sample versus the control in order to distinguish true bait interaction partners from background contaminants. TMT 6-plex permits analyzing triplicates of both control and specific sample in a single experiment; however, this method is associated with lower sensitivity as compared to iTRAQ 4-plex labeling (8).

Acknowledgments

The research leading to these results has received funding from the European Community's Seventh Framework Programme FP7/2007-2013 under grant agreement number KBBE-2009-222883. Horticultural and technical support has been provided by Karen Staples and Steve Price.

References

1. Osman K, Higgins JD, Sanchez-Moran E, Armstrong SJ, Franklin FC (2011) Pathways to meiotic recombination in Arabidopsis thaliana. New Phytol 190:523–544

2. Sanchez-Moran E, Mercier R, Higgins JD, Armstrong SJ, Jones GH, Franklin FC (2005) A strategy to investigate the plant meiotic proteome. Cytogenet Genome Res 109:181–189

3. Hollingsworth NM, Ponte L (1997) Genetic interactions between HOP1, RED1 and MEK1 suggest that MEK1 regulates assembly of axial element components during meiosis in the yeast *Saccharomyces cerevisiae*. Genetics 147: 33–42

4. Cheng CH, Lo YH, Liang SS, Ti SC, Lin FM, Yeh CH et al (2006) SUMO modifications control assembly of synaptonemal complex and polycomplex in meiosis of *Saccharomyces cerevisiae*. Genes Dev 20:2067–2081

5. Brassica rapa Genome Sequencing Project Consortium (2011) The genome of the mesopolyploid crop species *Brassica rapa*. Nat Genet 43:1035–1039

6. Ferdous M, Higgins JD, Osman K, Lambing C, Roitinger E, Mechtler K et al (2012) Inter-homolog crossing-over and synapsis in Arabidopsis meiosis are dependent on the chromosome axis protein AtASY3. PLoS Genet 8:e1002507

7. Herzog F, Peters JM (2005) Large scale purification of the vertebrate anaphase-promoting complex/cyclosome. Methods Enzymol 398:175–195

8. Pichler P, Köcher T, Holzmann J, Mazanek M, Taus T, Ammerer G et al (2010) Peptide labelling with isobaric tags yields higher identification rates using iTRAQ 4-plex compared to TMT 6-plex and iTRAQ 8-plex on LTQ orbitrap. Anal Chem 82:6549–6558

Chapter 22

Identifying Meiotic Mutants in *Arabidopsis thaliana*

Wayne Crismani and Raphaël Mercier

Abstract

Arabidopsis is a very powerful tool for understanding meiosis in plants with genetic approaches. We provide here a simple summary of the techniques used to test if a candidate gene has an essential meiotic function. These protocols require no specific prior knowledge and help eliminate easily avoided mistakes in the attribution of a meiotic function to your favorite gene.

Keywords Screen, Mutant, *Arabidopsis*, Meiosis

1 Introduction

The knowledge on the molecular mechanisms of plant meiosis has been increasing rapidly in the last decade (1, 2). While a number of species have contributed to the knowledge base, *Arabidopsis* is currently one of the most powerful plant model species by virtue of its small sequenced genome and the large array of tools, notably the huge collections of tagged mutants. The plant's small size and short life cycle make genetic analysis very efficient. Thanks to these features, a rapidly growing list of upwards of 50 genes with essential meiotic functions have been identified and characterized in *Arabidopsis*. The post-genomic era has created a huge number of avenues for meiotic research, which laboratories may wish to pursue. There are still many candidate genes that may have meiotic functions waiting to be discovered. Some possible sources of candidate genes include:

1. The literature. The workload of dissecting meiosis has been shared over a number of model species (3). A subset of meiotic genes is conserved evolutionarily at the sequence and/or structural levels, providing obvious targets for reverse genetics. However, most of the conserved genes have now been tested and characterized in *Arabidopsis* thanks to international efforts.

Wojciech P. Pawlowski et al. (eds.), *Plant Meiosis: Methods and Protocols*, Methods in Molecular Biology, vol. 990, DOI 10.1007/978-1-62703-333-6_22, © Springer Science+Business Media New York 2013

2. Transcriptomics and proteomics. A gene or protein expressed in flower buds at the right stage, but even better, in meiocytes, is a good candidate for having a meiotic function. This approach has proved successful in several cases (4–6). Now, it is likely to be even more useful with the high quality transcriptomic data on RNA isolated from *Arabidopsis* meiocytes (7–9).

3. Co-immunoprecipitation and yeast-2-hybrid interactome data. Proteins interacting with a protein known to have a meiotic function are also attractive candidates for characterization (10).

It is very likely that almost all of the genes whose disruption gives strong meiotic phenotypes have already been discovered in *Arabidopsis* thanks to a large forward genetic screen based on a dramatic meiotic defect giving a strong reduction in fertility (11). Duplicated genes, however, escape detection in such a screen (12). A growing number of genes are being discovered that have crucial roles during meiosis but do not exhibit easily detectable reduction in fertility (the targeted phenotype in meiotic forward genetic screens so far) when mutated. For example, a strong phenotype easily detected at the macroscopic level is a failure to repair meiotic double strand breaks (13–16). On the other hand, defects such as reduction in the number of crossovers as strong as 60% or a complete failure to undergo the second meiotic division are virtually undetectable at the macroscopic level (17). Detailed analyses at the microscopic level can easily reveal such defects. Consequently, it is very likely that there is a large number of genes with essential meiotic functions that have not been discovered yet as they do not show detectable phenotypes in forward screens performed so far. Hence, if a gene is suspected of having a meiotic function, but it is not currently annotated as having a meiotic function as such, a closer look at its respective mutants could be well worth the effort. Here, we present a protocol to test a gene, or a set of genes, for meiotic functions.

2 Materials

2.1 Pollen Staining

Alexander stain (18): ethanol (10% v/v), malachite green (0.01% w/v), distilled water (50% v/v), glycerol (25% v/v), phenol (5.5% v/v), chloral hydrate (5% w/v) (see Note 1), fuchsin acid (0.05% w/v), orange G (0.005% w/v), and glacial acetic acid (2% v/v). Add ingredients in order listed using appropriate protective gear.

2.2 Tetrad Analysis

Toluidine blue stain: dissolve toluidine blue in water at 0.1% (w/v).

2.3 Other Materials

1. Dissection needles and fine forceps.
2. Standard microscope slides and coverslips.

3. A dissection microscope.

4. A bright field microscope with a 40× dry objective.

3 Methods

3.1 Finding Insertions in the Gene of Interest

Several *Arabidopsis* mutant collections have been generated around the world providing more than 450,000 mapped insertions. Comprehensive online tools allow browsing resources throughout the genome. These tools are all quite similar but have different interfaces.

http://gbrowse.arabidopsis.org/cgi-bin/gbrowse/arabidopsis/

http://signal.salk.edu/cgi-bin/tdnaexpress

http://urgv.evry.inra.fr/projects/FLAGdb++/HTML/index.shtml

3.2 Choosing and Confirming the Insertions

1. Order at least three different mutant alleles for each gene (see Notes 2 and 3).

2. Preferentially select insertions between the ATG and stop codons (see Note 4).

3. Be aware of the genetic background of the plants (see Note 5).

4. Genotype at least 20 plants from each mutant line (see Notes 6 and 7).

3.3 Finding Meiotic Defects

Several simple techniques can be used, all with their advantages and disadvantages (Table 1).

3.3.1 Fruit Length Examination

In *Arabidopsis*, a strong reduction in the seed number (a downstream result of a meiotic fault) causes a reduction in the fruit length. Therefore, one of the first tests typically applied to a

Table 1

Characteristics of different mutant screening techniques presented in this chapter

	Fruit length	Pollen staining	Tetrad analysis	Chromosome spreads
Ease	++++	++	++	+
Specificity to meiosis	+	++	++++	++++
Sensitivity	+	+++	++	++++

"+++++" is the most positive and "+" is the least positive with respect to the title of the respective row.

suspected meiotic mutant is to look for a correlation between the mutant homozygous genotype (because T-DNA insertions are typically recessive) and the phenotype assessed by the fruit length and seed number (see Notes 2 and 8). However, other non-meiotic defects can also affect fruit length and seed number, therefore this test is not highly specific.

3.3.2 Pollen Staining

Meiotic defects frequently reduce pollen viability, which can be assessed by Alexander staining (18) (see Notes 2, 8, and 9).

1. Collect fresh buds from healthy plants at the stage when the flowers are just starting to open and the white tips of the petals can be barely seen.

2. Using dissection needles, remove the four big anthers and place them on a microscope slide.

3. Add a drop of Alexander stain.

4. Apply a coverslip. Gently press down the coverslip to help the stain enter the anthers. Be careful not to press so hard as to crush the pollen.

5. Leave the slide overnight. Observations are made using a compound microscope with a dry objective at a 40× magnification. Viable pollen grains will appear round, red and enveloped by a thin green layer. Nonviable pollen appears green, is usually smaller than wild-type pollen, and appears "deflated."

3.3.3 Tetrad Analysis

A male meiosis produces four spores embedded in a tetrahedral structure. Presence of abnormal numbers of spores or unbalanced spores is a direct sign of a meiotic defect (see Note 2).

1. Collect fresh buds from healthy plants at the stage when the bud measures approximately 0.5 mm (measurements are typically made at a 40× magnification using scales within the eyepiece of a dissection microscope—do not include the peduncle in the measurement).

2. Open the buds using dissection needles, remove the four big anthers, place them on a microscope slide.

3. Add a drop of Toluidine Blue stain and then apply a coverslip.

4. Still under the dissection microscope, using a pair of forceps, gently press down the coverslip until the anthers begin to break releasing cells.

5. Leave the slide for 5 min at room temperature to allow the dye to stain the cells.

6. Tetrads can be viewed with a compound microscope with a 40× dry objective. In wild type, because of the tetrahedral shape of the tetrad, often only three spores are visible in a focal plane. Adjusting the focus allows visualization of the fourth spore.

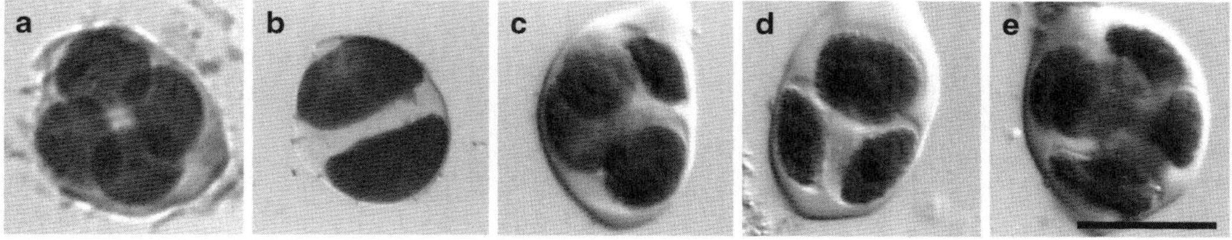

Fig. 1 Examples of tetrad analysis results. (**a**) Wild-type tetrad. (**b**) Dyads instead of tetrads, suggesting a possible defect in the first or second meiotic division (e.g., *osd1* (20)). (**c**) Unbalanced tetrad, suggesting possible mis-segregation of chromosomes at the first or second division (e.g., *Atmlh3* (17)). (**d**) Triads, suggesting a possible defect in the first or second division (e.g., *Atps1* (4)). (**e**) Polyads typically caused by either chromosome fragmentation and/or massive chromosome mis-segregation and often associated with very low levels of crossovers (*Atspo11-1* (21), *Atrad51* (13)). Scale Bar = 10 μm

Table 2
The relative sensitivity of each technique for detecting certain types of meiotic defects

	Fruit length	Pollen staining	Tetrad analysis	Meiotic spreads
Abolition of DSB formation	++++	++++	+++++	+++++
DSB repair defect	+++++	+++++	+++++	+++++
85% reduction in CO formation	++++	+++++	++++	+++++
60% reduction in CO formation	+	++++	+++	+++++
Absence of a division	+	++	+++++	+++
Premature loss of centromeric cohesion	++	++++	++	++++

"+++++" means "obviously detected," "++++" means "easily detected," "+++" means "detectable," "++" means "detectable with some experience" and "+" means "very hard to detect"

Confirmation of a meiotic defect using tetrad analysis definitely warrants further investigation to identify the nature of the defect. Since in wild-type plants tetrads are observed almost exclusively, presence of abnormalities at a frequency as low as 5% suggests a meiotic defect. Based on the type of abnormal structure(s) seen, one can begin to have an idea about the nature of the meiotic defect (Fig. 1).

3.3.4 Meiotic Chromosome Spreads

A powerful, but time consuming, confirmation of a meiotic defect is to conduct meiotic chromosome spreads (19). Preparing chromosome spreads is described in details in Chapter 1.

In Table 2, we present a summary of the effectiveness of the techniques presented above in detecting specific meiotic defects. These techniques can be used for forward as well as reverse genetic screens (see Note 10).

4   Notes

1. Chloral hydrate is regulated as a narcotic in some parts of the world, including the USA. Government permits may be required to purchase and/or use it.

2. Beware of chromosome translocations. At least 10% of T-DNA lines have massive translocations and likely another 10% have smaller chromosomal rearrangements. The T-DNA insertion process itself can cause translocations and, therefore, when a translocation occurs, it often is inherently linked to the T-DNA insertion site. Translocations can prevent correct analysis of the disruption of your gene. More specifically, one cannot be sure if any phenotype seen is due to a translocation or a disruption of the gene. This problem is particularly acute for meiosis research because translocations can lead to reduction in fertility, reduction in pollen viability, and chromosome segregation defects. Therefore, it is of extreme importance to check for the presence of translocations in your T-DNA lines and to subsequently exclude any lines with translocations from analysis. A rapid way to detect a translocation linked to a T-DNA insertion is to check the viability of pollen in a heterozygous plant (see Subheading 3.3.2). A translocation in a heterozygous state will have a dramatic negative effect on pollen viability because many pollen grains are lacking a part of their genome.

3. Approximately 20% of the insertion site predictions are not confirmed and hence we advise ordering a minimum of three mutant lines per gene, if available. Exercise care if multiple insertions appear in identical or even very similar positions in the genome as they often originate from PCR contamination, although one of the insertions is generally real. If this is the case, order all insertion lines that are available and expect to have only one allele.

    It is essential to have multiple alleles to confidently attribute a phenotype to the disruption of a specific gene. On average, T-DNA lines contain two T-DNA insertions as well as five other mutations that originate from the T-DNA insertion process but do not contain T-DNAs at the lesion site. To conclude that a phenotype is linked to the disruption of a gene, the same phenotype needs to be obtained in at least two independent mutant alleles. A complementation test should also be performed. If only one mutant allele is available for your favorite gene, a molecular complementation test using transformation with a wild-type version of the gene is required.

4. Typically, insertions 5′ of the ATG or 3′ of the stop codon do not disrupt the gene's function. Between the ATG and the

stop codon, any insertion, regardless of whether it is in an exon or intron, will typically disrupt the function of the gene.

5. Most mutant collections (SALK, SAIL, GABI, Wisc, and SK) were produced in the Columbia ecotype background, allowing direct comparison of phenotypes between mutants from these collections. However, some collections are in different backgrounds, which can be useful in certain circumstances. The CSHL and JIC collections are in the Landsberg background, FLAG is in the Wassilewskija background, and Riken is in the Nossen background.

6. The genealogy of the seeds received from stock centers is often unclear (except for GABI where the seeds are harvested from individual sister plants). Consequently, we advise genotyping at least 20 plants. We also advise against analyzing the segregation of the mutation until you have harvested seeds from your own genotyped plants.

7. Primers specifically designed for your lines are easily generated with http://signal.salk.edu/tdnaprimers.2.html. Exercise care to use the correct primer for the insertion unique to each mutant collection.

8. The first five flowers of each stem often contain a lot of dead pollen in wild-type plants. Accordingly, the first five fruits are often short, and should not be taken into account to assess fertility. Be patient and wait for a few days more.

9. Most meiotic defects have a comparable effect on male and female. However, male meiosis is typically studied because of a larger number of male meiocytes, which are easier to access than female meiocytes. However, female meiosis is the more limiting factor for fertility, as pollen grains are in large excess.

10. All of these techniques can be used for a forward genetic screen if a mutant collection is available. Large forward genetic screens have identified many *Arabidopsis* meiotic mutants based on a strong reduction in fertility. Hence, screening for this phenotype is unlikely to provide new unique mutants. However, screens based on other more subtle phenotypes listed here would identify mutants not found so far. T-DNA mutant collections can be used for this purpose because the tagged mutations make identification of the insertion relatively easy. However, the affordability of deep sequencing is increasing the feasibility of working with techniques that use a higher rate of mutagenesis such as EMS. Use Tables 1 and 2 to optimize your screen according to the desired phenotype. More sophisticated approaches, such as suppressor screens or screens for altered recombination rates between linked markers can also be designed.

References

1. Osman K, Higgins JD, Sanchez-Moran E, Armstrong SJ, Franklin FCH (2011) Pathways to meiotic recombination in *Arabidopsis thaliana*. New Phytol 190:523–544

2. Mercier R, Grelon M (2008) Meiosis in plants: ten years of gene discovery. Cytogenet Genome Res 120:281–290

3. Gerton JL, Hawley RS (2005) Homologous chromosome interactions in meiosis: diversity amidst conservation. Nat Rev Genet 6: 477–487

4. d' Erfurth I, Jolivet S, Froger N, Catrice O, Novatchkova M, Simon M et al (2008) Mutations in AtPS1 (Arabidopsis thaliana parallel spindle 1) lead to the production of diploid pollen grains. PLoS Genet 4:e1000274

5. Wijeratne AJ, Chen C, Zhang W, Timofejeva L, Ma H (2006) The *Arabidopsis thaliana PARTING DANCERS* gene encoding a novel protein is required for normal meiotic homologous recombination. Mol Biol Cell 17: 1331–1343

6. Osman K, Sanchez-Moran E, Mann SC, Jones GH, Franklin FCH (2009) Replication protein A (AtRPA1a) is required for class I crossover formation but is dispensable for meiotic DNA break repair. EMBO J 28:394–404

7. Libeau P, Durandet M, Granier F, Marquis C, Berthomé R, Renou JP et al (2011) Gene expression profiling of *Arabidopsis* meiocytes. Plant Biol 13:784–793

8. Yang H, Lu P, Wang Y, Ma H (2011) The transcriptome landscape of *Arabidopsis* male meiocytes from high-throughput sequencing: the complexity and evolution of the meiotic process. Plant J 65:503–516

9. Chen C, Farmer AD, Langley RJ, Mudge J, Crow J, May GD et al (2010) Meiosis-specific gene discovery in plants: RNA-Seq applied to isolated *Arabidopsis* male meiocytes. BMC Plant Biol 10:280

10. Dreze M, Carvunis A-R, Charloteaux B, Galli M, Pevzner SJ, Tasan M et al (2011) Evidence fornetwork evolution in an *Arabidopsis* interactome map. Science 333:601–607

11. De Muyt A, Pereira L, Vezon D, Chelysheva L, Gendrot G, Chambon A et al (2009) A high throughput genetic screen identifies new early meiotic recombination functions in Arabidopsis thaliana. PLoS Genet 5:e1000654

12. Higgins JD, Sanchez-Moran E, Armstrong SJ, Jones GH, Franklin FCH (2005) The *Arabidopsis* synaptonemal complex protein ZYP1 is required for chromosome synapsis and normal fidelity of crossing over. Genes Dev 19:2488–2500

13. Li W, Chen C, Markmann-Mulisch U, Timofejeva L, Schmelzer E, Ma H et al (2004) The *Arabidopsis AtRAD51* gene is dispensable for vegetative development but required for meiosis. Proc Natl Acad Sci U S A 101:10596–10601

14. Puizina J, Siroky J, Mokros P, Schweizer D, Riha K (2004) Mre11 deficiency in Arabidopsis is associated with chromosomal instability in somatic cells and Spo11-dependent genome fragmentation during meiosis. Plant Cell 16:1968–1978

15. Schommer C, Beven A, Lawrenson T, Shaw P, Sablowski R (2003) AHP2 is required for bivalent formation and for segregation of homologous chromosomes in *Arabidopsis* meiosis. Plant J 36:1–11

16. Kerzendorfer C, Vignard J, Pedrosa-Harand A, Siwiec T, Akimcheva S, Jolivet S et al (2006) The *Arabidopsis thaliana* MND1 homologue plays a key role in meiotic homologous pairing, synapsis and recombination. J Cell Sci 119:2486–2496

17. Jackson N, Sanchez-Moran E, Buckling E, Armstrong SJ, Jones GH, Franklin FCH (2006) Reduced meiotic crossovers and delayed prophase I progression in AtMLH3-deficient *Arabidopsis*. EMBO J 25:1315–1323

18. Alexander M (1969) Differential staining of aborted and nonaborted pollen. Biotech Histochem 44:117–122

19. Ross KJ, Fransz P, Jones GH (1996) A light microscopic atlas of meiosis in Arabidopsis thaliana. Chromosome Res 4:507–516

20. d' Erfurth I, Jolivet S, Froger N, Catrice O, Novatchkova M, Mercier R (2009) Turning meiosis into mitosis. PLoS Biol 7:e1000124

21. Grelon M, Vezon D, Gendrot G, Pelletier G (2001) AtSPO11-1 is necessary for efficient meiotic recombination in plants. EMBO J 20:589–600

Wojciech P. Pawlowski et al. (eds.), *Plant Meiosis: Methods and Protocols*, Methods in Molecular Biology, vol. 990,
DOI 10.1007/978-1-62703-333-6, © Springer Science+Business Media New York 2013

Printed by Printforce, the Netherlands